Understanding GIS

GIS

An ArcGIS Project Workbook

Christian Harder | **Tim Ormsby** | **Thomas Balstrøm**

Published by Esri Press, Redlands, California

Ask for Esri Press titles at your local bookstore or order by calling 1-800-447-9778. You can also shop online at www.esri.com/esripress. Outside the United States, contact your local Esri distributor.

Esri Press titles are distributed to the trade by the following:

In North America:

Ingram Publisher Services

Toll-free telephone: 1-800-648-3104

Toll-free fax: 1-800-838-1149

E-mail: customerservice@ingrampublisherservices.com

In the United Kingdom, Europe, Middle East and Africa, Asia, and Australia:

Eurospan Group

3 Henrietta Street

London WC2E 8LU

United Kingdom

Telephone: 44(0) 1767 604972

Fax: 44(0) 1767 601640

E-mail: eurospan@turpin-distribution.com

Foreword

Over twenty years ago I was involved with a book project here at Esri. We produced and published a software workbook called *Understanding GIS: The ARC/INFO Method*. We were just a group of people at this little GIS software company in Redlands, CA who saw that the users of our Arc/Info software needed better guidance on how to use these tools for real analysis projects. Our original goal had been to write a book called *Getting Started with ArcInfo*. 500+ pages later we realized that we had described a methodology for doing GIS projects and needed to change the title.

The key idea was to set up a problem in the book, provide the data, and then let the students work through the entire process. Little did we know at the time that our book would become a worldwide bestseller and inspire a generation of technically savvy geographers.

Fast forward to 2011. GIS has evolved to keep pace with a changing technology landscape. There have been huge innovations in geospatial data models and user interfaces; GIS now lives and breathes on the web. At lot has changed. Yet the need for our users to understand how to organize and think about a GIS project persists. This book, *Understanding GIS: An ArcGIS Project Workbook* revives the spirit and method of the original that inspired it. I'm pleased to say that the new edition carries the torch while also integrating the new paradigm. It is my hope that it will inspire a new generation of GIS professionals and practitioners.

Clint Brown
Director of Software Products
Esri, Redlands, California

Table of contents

Preface

Background

In the late 1980s, Esri published a software workbook called *Understanding GIS: The ARC/INFO Method*. This book was the first to offer a practical, project-oriented introduction to a commercial GIS software product, and it became popular as both a self-study tutorial for working professionals and as a lab manual in college classrooms.

ArcGIS Desktop 10 is different from the Arc/Info software the original book was written for. Computers and operating systems are different, too. Ongoing development by Esri has extended the software's capabilities and reworked its architecture to keep pace with advances in technology. Not the least of these advances is the sharing of geographic data and applications on the web. When the original book was published, the World Wide Web was still a couple of years in the future. In addition to new GIS functionality, there have also been significant changes to GIS data models, data storage, and user interfaces. All these changes notwithstanding, the underlying geographic approach (what was then called "The Arc/Info Method") remains basically the same. That approach can be summarized as follows: frame the problem, explore the study area, prepare the data, perform the analysis, and present the results. The same method is followed in this new book.

This book isn't a new edition of the original—the original is part of software history. This is a new book from start to finish. It draws inspiration from the original, but incorporates the latest developments to ArcGIS and reflects the world in which GIS is practiced today.

Description

Understanding GIS: An ArcGIS 10 Project Workbook is a tutorial designed around a single multifaceted problem: that of finding a suitable location for a park by the Los Angeles River in Los Angeles, California. Using real spatial data (from the City of Los Angeles and other providers) and realistic requirements, you'll complete all the essential phases of a GIS analysis project from planning, to execution, to follow-up.

Goals

One goal of this book is to help you become a proficient user of ArcGIS Desktop 10 software. Another goal, ultimately more important, is to teach the geographic approach to problem-solving. GIS students are often frustrated because software tools are presented in contexts that don't make clear how the tools relate to one another or how they serve larger purposes. This book incorporates software functionality in a meaningful process of analytic thinking.

Of course, GIS has many uses that are not analytical—data management and cartography to name a couple. Even within the general category of analysis, there are different kinds of problems requiring different tools and strategies. This book is not all things GIS. It's not all things ArcGIS, either. There are many aspects of the software that aren't relevant to our problem and aren't explored in the book. Nevertheless, the book covers a lot of ground. When you're done, you should have a good working knowledge of the software and a strong sense of how to use it productively on your own.

Audience

People come to GIS with varying background knowledge and experience. We loosely classify this book as intended for "ambitious beginners." If you have no prior experience with GIS or any of its wellsprings (geography, cartography, earth science, and computer database technology among them), you may find it more challenging than someone with exposure to these areas. On the other hand, thanks to GPS and household Internet mapping applications, location-based technology is becoming ever more familiar. Many things that needed explanation a few years ago have now merged into the background of common technical savvy.

This book is mainly written for

- College-level or graduate students in GIS-related disciplines who want to learn ArcGIS Desktop software and GIS practice
- College and university teachers who want a classroom lab manual to supplement their GIS instruction
- New GIS professionals who want to strengthen or expand their knowledge of ArcGIS Desktop 10 software
- Professionals in other technical fields who want cross-training in GIS
- Current ArcGIS Desktop users transitioning from earlier software versions to version 10

There are no prerequisites for the book, and we've tried to write it so it will work for anyone with a serious interest in learning GIS. During the course of development and review, we found that many experienced users found things of value in it, too. If you're brand new to GIS and are using the book outside a classroom or support group, be sure to take full advantage of the book's resource center—especially the GIS TV videos.

Structure

As mentioned, the book follows a single project, so it's best to work through it from start to finish. Lesson results are provided in the event that you can't do every lesson or run into trouble.

The book has nine lessons, each divided into two or more exercises of varying length. Lesson 9 is only introduced in the book: the exercises for this Lesson only need to be downloaded from the Understanding GIS Resource Center website.

Exercises are fully scripted, so you won't encounter gaps in instruction, but frequently repeated operations aren't spelled out in detail every time. For example, after you've clicked a toolbar button a few times, we won't keep showing pictures of it. Likewise, after you've opened a dialog box a few times, we won't keep telling you how to open it.

Conceptual information is supplied as needed, either directly in the text or in call-out boxes. A few complex topics have a page to themselves. Figures, mostly screen captures, are common throughout the book. Their main purpose is to show you the correct state of your software at a particular point. Some figures are annotated to emphasize settings that need attention or to help interpret a result.

Appendix A discusses assumptions and complications lying behind the analysis. Appendix B is the data license agreement. The book's index will help you find your way back to important concepts, operations, and tools.

The exercises were created on computers running the Microsoft® Windows XP® operating system. They were tested on machines running Windows XP and Windows 7®. Where exercise instructions involve operating system paths, we've used Windows XP names, but tried to note differences in Windows 7. All images that are screen captures of software reflect a default Windows XP configuration.

Resources

Understanding GIS Resource Center. The book's online support site is the Understanding GIS Resource Center:

http://resources.arcgis.com/Understanding-GIS

Come here to get lesson results and download additional exercises and videos.

What is GIS and what is ArcGIS?

Before you start this book, especially if you are new to GIS, we strongly recommend that you read the section "What is GIS?" in the Essentials Library of the ArcGIS 10 Desktop help. "What is GIS?" explains and illustrates foundation concepts, your knowledge of which this book may sometimes take for granted. Also useful is the next section of the Essentials Library, "What is ArcGIS?" which describes the products and technologies that make up the ArcGIS software system. You will find links to these sections of the desktop help at the Understanding GIS Resource Center (see URL above).

Lesson results. The book's exercises are cumulative, with the results of one exercise defining the starting point of the next. For this reason, your results at the end of every lesson have to be right. The lessons include many screen images as visual confirmations of progress, so by the end of an exercise, you should know whether you got the correct results. If you did, you can carry them forward. If you have problems, or if you skip an exercise, you can get result data and maps for any lesson. Results are available at the Understanding GIS Resource Center.

Additional exercises. The exercise instructions for Lesson 9, "Share results online," need to be downloaded from the Understanding GIS Resource Center. This lesson relies on web-mapping applications and websites that are updated on different schedules from ArcGIS Desktop software. Locating the lesson online insures that it will stay current.

Videos. Every exercise has a narrated how-to video that shows the exercise being done in real time. You may find it helpful to watch a video before doing an exercise; the videos may also be a good refresher later. These GIS TV videos can be accessed from the Understanding GIS Resource Center.

Other resources. The ArcGIS Desktop Help System, installed with the software and accessible online, provides comprehensive descriptions of software concepts and tools. Additional online resources such as blogs, forums, map galleries, videos, user communities, and access to technical support and training can be located at this website:

http://resources.arcgis.com

Disclaimer

The data used in this book is real. So are the efforts of the City of Los Angeles and several interest groups to improve the environmental quality of the Los Angeles River and its surroundings. The GIS project in this book, however, was developed entirely at Esri. For the sake of a story, we pretend that the project was sponsored by the Los Angeles City Council, and throughout the book there are references to what the "city council" wants or expects. In fact, neither the book nor the project have any affiliation with the city beyond permission to use its GIS data. Likewise, there is no affiliation with the Los Angeles River Revitalization Project or with any Los Angeles River advocacy organization.

Acknowledgments

This book would not have been possible without the cooperation of the City of Los Angeles, Department of Public Works, Bureau of Engineering. Special thanks to Randy Price and Ann-Kristin Karling of the Bureau's Mapping Division, and to City Engineer Gary Lee Moore, for giving us access to the city's data. We thank the City of Los Angeles Department of Recreation & Parks for providing its parks data. We note that land parcels and attributes are maintained by the Los Angeles County Assessor's Office.

The idea for the GIS project in this book was inspired by the Los Angeles River Revitalization Master Plan (http://www.lariverrmp. org). Images from the master plan are used in Lesson 1 by permission.

Many people reviewed the book in whole or in part, tested exercises, and gave advice or help. Thanks to Mamata Akella, Richard Alden, Matthew Baker, John Berry, Janel Day, Drew Flater, Charlie Frye, Koushik Hajra, Dale Honeycutt, Melita Kennedy, Deane Kensok, Andy Mitchell, Eileen Napoleon, Brian Parr, Brandy Perkins, Tom Proctor, Veronica Rojas, Ryan Theis, Lindsay Thomas, Vin Thomas, Natalie Vines, and Molly Zurn.

Thanks to Joyce Frye for copyediting the manuscript.

Thanks to Veronica Rojas for producing the GIS TV videos.

Thanks to Catherine Jones for management support.

Thanks to Peter Adams, David Boyles, Donna Celso, Brian Harris, Jennifer Hasselbeck, and Jay Loteria at Esri Press and to Cliff Crabbe for patiently shepherding the book through production.

And, most of all, a huge thanks to Clint Brown for supporting this project through thick and thin.

Technical requirements

For this book, your computer needs the following components:

- 2.2 GHz CPU with an Intel Core Duo, Pentium 4, or Xeon processor, 2GB of RAM and at least 2.4 GB free disk space with up to 50 MB of disk space in the Windows System directory.
- Microsoft Windows XP, Vista, or Windows 7 operating system (with the most recent service packs).
- Microsoft .NET 3.5 SP1 must be installed prior to installing ArcGIS Desktop 10.
- ArcGIS Desktop 10 software, ArcInfo license level (provided).
- Additional 796 MB free disk space for installing the exercise data (exercise data provided).
- Either Microsoft Internet Explorer Version 8 (or newer) or Mozilla Firefox Version 2 (or newer).
- An Internet connection.
- Visit http://www.esri.com/AG10systemrequirements for more information.

Installing ArcGIS Desktop 10 software

This book provides a 90-day trial version of ArcGIS Desktop 10 software—ArcInfo single-use license—which can be downloaded at **www.esri.com/90daytrial**. Use the code printed on the inside back cover of this book to authorize your 90-day trial, and then follow the onscreen instructions to complete the software install. This is a fully functional, but nonrenewable, 90-day license.

If you already have ArcGIS Desktop 10 with an ArcInfo license on your computer, you can skip this step.

Required software version

This book requires ArcGIS Desktop 10 software with an ArcInfo license. Earlier software versions are not compatible with exercise data and do not operate as described in the exercises.

Installing the project data

Follow the steps below to install the exercise data.

1) Put the data DVD in your computer's DVD drive. A splash screen will appear.

2) Read the welcome, then click the Install Exercise Data link. This launches the InstallShield Wizard.

3) Click Next. Read and accept the license agreement terms, then click Next.

4) Accept the default installation folder or click Browse and navigate to the drive or folder location where you want to install the data.

5) Click Next. The installation will take a few moments.

6) Click Finish. The exercise data is installed on your computer in a folder called C:\UGIS (or \UGIS in the folder where you installed the exercise data).

1 Frame the problem and explore the study area

THE VOLATILE LOS ANGELES RIVER

is the reason America's second-largest city was founded in its present southern California location by Spaniards in 1781. (The area was originally settled by the Gabrielino Native American Tribe thousands of years earlier). Its water was tapped for drinking and irrigation, and a new city spread out from the river across the coastal plain. By the turn of the 20th century, the river was surrounded by a thriving urban center. Every few decades, raging floods would crest the banks at various points, submerging entire neighborhoods. After the historic floods of 1938 that claimed over 100 lives and washed out bridges from Tujunga Wash to San Pedro, city leaders had seen enough. By 1941 the U.S. Army Corps of Engineers began to straighten, deepen, and reinforce the once wild waterway. Much of its length was eventually lined in concrete, and the river was more or less tamed.

Today, the City of Angels—home to nearly 4 million people—is a vibrant world center of business and culture. Running straight through the heart of the city, the Los Angeles River now serves ably as a flood-control channel. Sadly, this once-bucolic waterway that was so instrumental in the formation of the city became known as something ugly and marginal. Mile after mile of angled concrete appealed only to graffiti artists and filmmakers, and save for the occasional televised rescue of some hapless Angeleno swept away by a winter storm-fed torrent, the river remained a part of the city ignored by most. The negative perception has stuck with the neglected river for decades.

But in recent years, as the city has densified and much of southern California's wildlands have been appropriated for development, new attention has focused on the river corridor and the scattered pockets of open space that line its length. While it must always serve its important flood control function, the river and adjacent lands are increasingly recognized as under-utilized, providing opportunities for regreening and psychic restoration for human beings in an overbuilt city. Adventuresome and resourceful citizens have discovered peaceful pockets of sanctuary along the river and made these places their own. A vital and concerned activist community has raised awareness of the river and pushed for its beautification and redevelopment.

Photo by Los Angeles Times Staff Photographer
Copyright © 1938, Los Angeles Times
Reprinted with permission

In 2005, city officials launched a major public works project focused on the human dimensions of the river. A landmark study, the Los Angeles River Revitalization Master Plan, demonstrated the significant potential to improve quality of life for citizens living near the river corridor through wise redevelopment. Mayor Antonio Villaraigosa said at the time: "We have an opportunity to create pocket parks and landscaped walkways...to create places where children can play and adults can stroll."

According to city leaders, "The Plan provides a 25- to 50-year blueprint for transforming the City's 32-mile stretch of the river into an 'emerald necklace' of parks, walkways, and bike paths, as well as providing better connections to the neighboring communities, protecting wildlife, promoting the health of the river, and leveraging economic reinvestment."

While the 2005 master plan identified some of the most obvious areas for large-scale regional redevelopment along the river, it stopped short of identifying smaller (and more affordable) neighborhood projects; that work would require a more involved study. With thousands of land parcels strung out along the river, identifying the best places for park development is like looking for that proverbial needle in a haystack. Many factors come into play, among them current land use, demographics, and accessibility.

In the years since the Plan's completion, the City has created a website that encourages people to learn about (and participate in) the latest developments related to its landmark resource. The website at www.LARiver.org contains links to many resources about the river and its watershed, including scientific studies and recreational opportunities. If city leaders can find the resources and a motivated citizenry keeps up the pressure, there will be a renaissance transforming growing stretches along the river that become real versions of these artists' renderings.

All images on this page courtesy of The City of Los Angeles, LA River Revitaliaztion Master Plan

Here's where you pick up the thread in this book. You'll use the city's real need for river redevelopment as a launching point for a park siting analysis using a geographic information system (GIS). A GIS is ideal for this type of decision-making because it allows you to analyze large amounts of data in a spatial context. In this book, you'll spend a lot of time with Esri's ArcGIS® Desktop software, and by the end you'll have completed a project from start to finish. Along the way, you'll gain an excellent grasp of what a GIS can do.

You'll be assuming the role of a GIS analyst for the City of Los Angeles. So what exactly does a GIS analyst do and how is that job different from other jobs that also use GIS software? The table below defines some of the various roles that a typical GIS operation might establish to accomplish their work.

GIS Roles	
Editors	People who create, update, and correct spatial data and its attributes (statistical and descriptive information).
Cartographers	People who make and publish maps and solve information design problems.
Analysts	People who query and process geographic data to solve analytical problems.
Programmers	People who implement custom GIS functionality by developing scripts and applications for specific procedures.
Managers	People who oversee staffing and equipment, database design, workflow, new technology and data acquisition.

The central work in this book is analytical. Your main focus will be on using ArcGIS tools and methods to find the most park-suitable land within a study area, but there is preparatory work to do before the analysis proper, and there are results to interpret and present afterwards. This book has two goals. One is to present a comprehensive approach to geographic problem solving. We want to help you develop skills, habits, and ways of thinking that will be useful in projects other than this one. The second goal is to teach you how to use ArcGIS Desktop 10 software. These goals are mostly complementary. ArcGIS is a big system, however, and it wouldn't be realistic to try to cover all it can do in a single book. Our principle has been to teach the software in the service of the project and not otherwise. You'll delve into many aspects of ArcGIS Desktop—editing, modeling, and cartography among them—but there are other aspects we won't use, or will only touch lightly, because they aren't strictly relevant to our needs. We might say (with apologies to Waldo Tobler) that everything in a GIS is related to everything else, but some things are more closely related to analysis than others.

Frame the problem

The first step in the geographic approach to problem-solving is to frame the problem. What that means, first of all, is coming up with a short statement of what it is you want to accomplish. We want to find a suitable site for a park near the Los Angeles River.

Once you have the statement, you can begin to tease out its ambiguities. What factors make a site "suitable"? Fortunately for us, the city council has already established a concise and fairly specific set of guidelines. They want a park to be developed:

- On a vacant parcel of land at least one acre in size
- Within the Los Angeles city limits
- As close as possible to the Los Angeles River
- Not in the vicinity of an existing park
- In a densely populated neighborhood with lots of children
- In a low-income neighborhood
- Where as many people as possible can be served

This limits the scope of our inquiry, but it's far from a complete breakdown of the problem. Some of the guidelines are specific, but others are vague. Familiar concepts are sometimes the hardest to pin down. For example, what income level should count as "low income"? How are the boundaries of a "neighborhood" established? We can't solve the big problem until we've solved the little problems buried inside it. Usually, however, it's not possible to address (or even foresee) all the little problems ahead of time.

Data exploration influences the framing of the problem. Do you have income data on hand? If so, is it for individuals or households? Is it average, median, or total? To the extent that the questions themselves are indefinite (what is low income? how should it be measured?), the data you have available will help shape the answer.

Analysis influences the framing of the problem. Given that we want a one-acre tract of vacant land, what do we do about adjacent half-acre lots? Is there a tool to combine them? If so, does it have undesired side effects, such as loss of information? Our data-processing capabilities (and our knowledge of them) may determine how we define a "one-acre parcel."

Even the results of an analysis influence the framing of the problem. Suppose, after having carefully defined the guidelines, we run the analysis and don't find any suitable sites. Do we make a report to the city council that there's just no room for a park anywhere? More likely, we change some of our definitions and run the analysis again.

Framing the problem, therefore, is an ongoing process, one that will occupy us through much of the book.

Lesson One roadmap

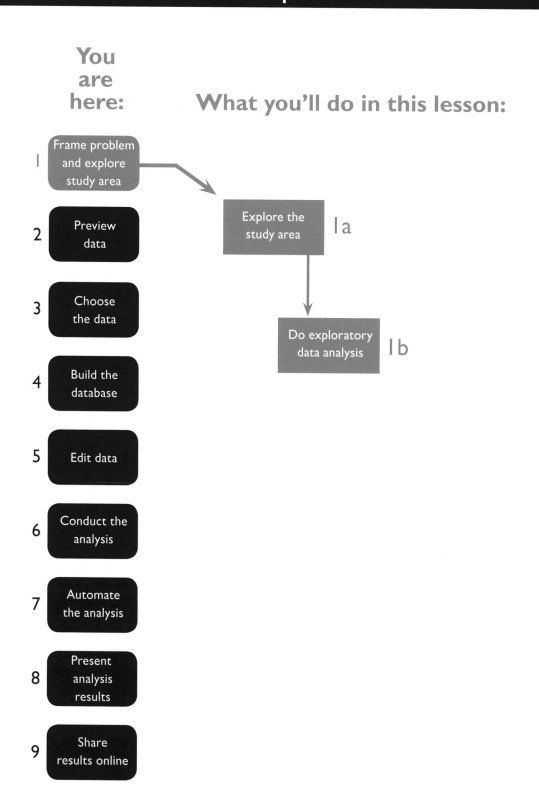

You are here:

What you'll do in this lesson:

1. Frame problem and explore study area

2. Preview data

3. Choose the data

4. Build the database

5. Edit data

6. Conduct the analysis

7. Automate the analysis

8. Present analysis results

9. Share results online

Explore the study area — 1a

Do exploratory data analysis — 1b

Exercise 1a: Explore the study area

In this exercise, we'll get to know the Los Angeles River and its surrounding area with maps and data. At the same time, we'll learn the basics of working with ArcMap: how to navigate a map, how to add and symbolize data, and how to get information about map features.

1) **Start ArcMap.** We'll open the ArcMap application.

Ⓐ Click the Windows Start button, then choose All Programs → ArcGIS → ArcMap 10.

The application opens with the ArcMap - Getting Started dialog box in the foreground (Figure 1-1).

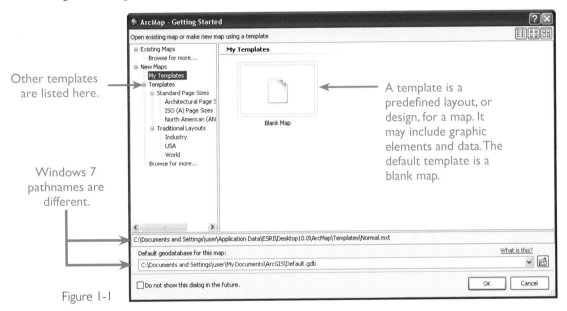

Other templates are listed here.

Windows 7 pathnames are different.

A template is a predefined layout, or design, for a map. It may include graphic elements and data. The default template is a blank map.

Figure 1-1

Ⓑ In the left-hand window, if necessary, click My Templates.

Ⓒ Click the Blank Map icon to select it, then click OK.

A blank map document opens.

2) **Add a basemap layer.** A basemap is a layer of reference geography that forms a backdrop to your map.

Ⓐ On the Standard toolbar at the top of the application window, locate the Add Data button ➕▾.

Ⓑ Click the small drop-down arrow next to the button.

▷ If you click the icon instead of the arrow, an Add Data dialog box will open. Close this dialog box and try again.

Ⓒ On the drop-down menu, choose Add Basemap.

D In the Add Basemap window, click *Streets* to select it (Figure 1-2), then click Add.

Figure 1-2

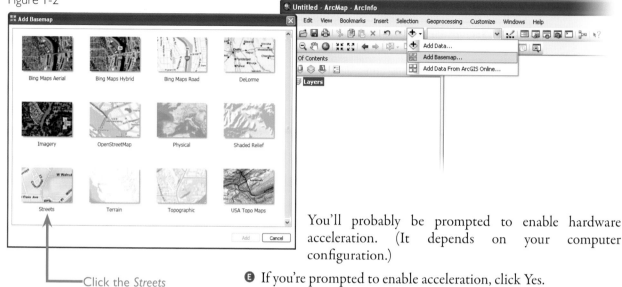

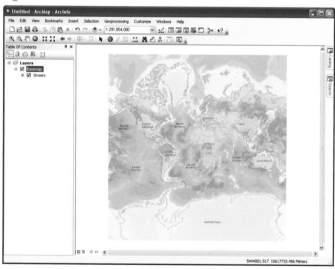

Click the *Streets* basemap

You'll probably be prompted to enable hardware acceleration. (It depends on your computer configuration.)

E If you're prompted to enable acceleration, click Yes.

This book relies on map services (remotely-hosted geographic data, like these basemaps) because that is how GIS is done today. One side effect is that there can be complications due to servers, Web and network connections, and the processing power of your computer. Therefore, operations in some cases may be slower than you anticipate. Be reasonably patient as you work, and check the Understanding GIS Resource Center (see page xii) for updated information if performance issues persist.

The *Streets* basemap layer is added to your map document (Figure 1-3). It has an entry in the Table of Contents window on the left.

Figure 1-3

At the top of the application is the main menu. Underneath it are two toolbars. The Standard toolbar has generic functions (like New, Save, Print, Copy, Paste) and buttons to open other windows. The Tools toolbar has navigation tools (like Zoom In, Zoom Out, and Pan) and other common tools for doing things with maps. There are lots of other toolbars besides these, and we'll use a couple of them in this lesson.

On the right side of the application are two tabs called Catalog and Search. These tabs are hidden windows. We'll use both windows a lot in the book, although we won't get to the Search window for a few more lessons.

Toolbars and windows can be moved around and resized. Figure 1-3 reflects the default arrangement, but your interface may look different if you've already been using the software.

3) Zoom in to southern California. We'll zoom in to our area of interest.

Ⓐ On the Tools toolbar, click the Zoom In tool 🔍. In the map window, click and drag to draw a box around Southern California, as shown in Figure 1-4.

▷ Your box doesn't have to match exactly.

Ⓑ Use the Zoom In tool again to get closer. When you see city names and major roads, use the Pan tool 🖐 to center the view on the Los Angeles area.

Ⓒ Keep zooming in (try the Fixed Zoom In button, 🔆 too) until you can easily distinguish cities, freeways, and landmarks like parks and airports.

Ⓓ Use the Pan tool and the various zoom tools (including Zoom Out 🔍 and Fixed Zoom Out ⛶) until your view more or less matches Figure 1-5.

You probably noticed that no streets were visible at the global scale, and that as you kept zooming in, more and more detail appeared. This is because the basemap is a multiscale map: really a set of maps which turn themselves on and off to display features and symbology that are appropriate to your map scale.

Figure 1-5 shows greater Los Angeles, an area that includes hundreds of incorporated communities and nearby cities. They're shown here as discrete points, but in reality they make up a contiguous urban patchwork, stitched together by freeways, and hemmed by mountains and the sea.

Figure 1-4

Draw a box around the area you want to zoom to

Figure 1-5

Basemap layers

Basemap layers show reference geography like street maps, imagery, topography, and physical relief. The ones available from the Add Basemap window in ArcMap® are remotely hosted map services that you can navigate, view, and use as backdrops to other data. Basemaps are stored at multiple scales, so that as you zoom in or out you see different amounts of detail. As you navigate a basemap, the various pieces of it (called tiles) that compose your current view are stored locally on your computer in a so-called display cache. When you zoom or pan to a new area, the map may be a little slow to draw, but any place you return to will redraw quickly because the data comes from your cache, not from the remote server.

Figure 1-6

4) Add a layer of project data. On top of the basemap, we'll add a layer from the supply of data that has been put together for this project and is stored on your computer. In Lesson 2, we'll talk more about where this data comes from, and how you acquire data of your own.

A On the Standard toolbar, click the Add Data button ✚. This time, click the icon itself, not the drop-down arrow.

The Add Data dialog box opens to a default "home" location in your MyDocuments\ArcGIS folder. (In Windows 7, the folder is Documents\ArcGIS.) We don't have any data here. To navigate to our data folder, we have to make a folder connection.

B In the row of buttons at the top of the Add Data dialog box, click the Connect To Folder button 🖳.

C In the Connect To Folder dialog box, navigate to C:\UGIS (Figure 1-6), then click OK.

D In the Add Data dialog box, navigate as follows:
- Double-click the ParkSite folder to open it.
- Double-click the Source Data folder.
- Double-click ESRI.gdb.
- Double-click Boundary.

We'll discuss the nature of data structures like "ESRI" and "Boundary" in the topic *Representing the real world as data* in Lesson 2 (page 63). For now, we just want to dig down to our data.

E In the Add Data dialog box, click *City_ply* to select it (Figure 1-7), then click Add.
- ▷ Or just double-click *City_ply* from the list.

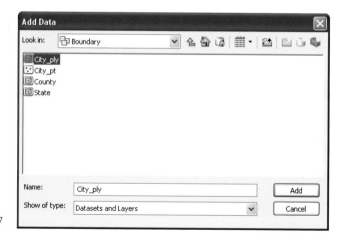

Figure 1-7

A layer of city boundaries is added to the map (Figure 1-8). Each city in the layer is called a feature. These features are polygons, which are one of the three basic shapes with which geographic objects are represented in a GIS. (The others are lines and points.)

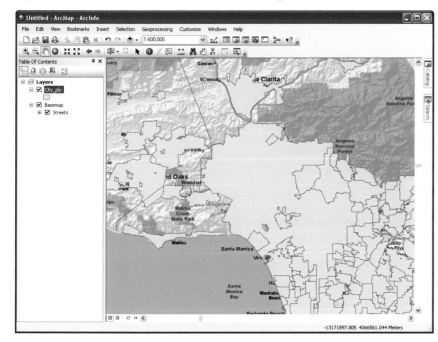

The color of the layer is assigned at random, so yours may be different.

Figure 1-8

In the Table of Contents, the order of entries matches the drawing order of layers in the map. *City_ply* is listed above *Basemap,* and on the map, the cities cover the basemap. You can control a layer's visibility with its check box in the Table of Contents.

F In the Table of Contents, click the check box next to *City_ply*.

The layer turns off.

G Click its check box again to turn the layer back on.

5) Set layer properties. Every layer has properties you can set and change. For example, you just changed the visibility property of the *City_ply* layer.

A In the Table of Contents, right-click the color patch underneath the *City_ply* layer name.

A color palette opens (Figure 1-9). Moving the mouse pointer over any color square shows its name as a tool tip.

B On the color palette, click any color you like to change the layer color.

On the map, the cities redraw in the color you chose.

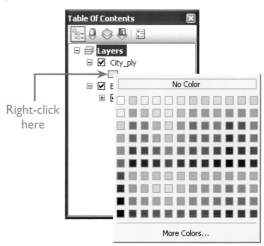

Right-click here

Figure 1-9

C In the Table of Contents, right-click the *City_ply* layer name to open its context menu. At the bottom of the menu, choose Properties.

The Layer Properties dialog box opens. This is where you access the full set of properties for a layer.

D If necessary, click the General tab (Figure 1-10).

Each tab along the top has different layer properties. We'll work with lots of layer properties in this book, but not all of them.

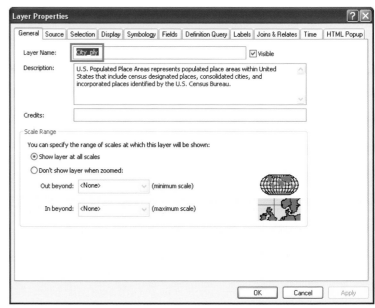

Figure 1-10

The layer's name, *City_ply,* is one of its properties. This name is cryptic (it stands for "city polygons") and unattractive, so let's change it.

E In the Layer Name box, delete the name and type **Cities** instead. Click Apply.

The name is updated in the Table of Contents.

F In the Layer Properties dialog box, click the Source tab.

This tab shows technical information about the layer, including the path to the data on your computer (Figure 1-11).

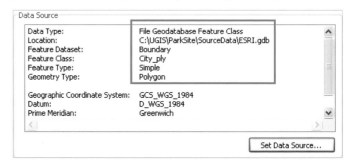

Figure 1-11

Renaming the layer in the map document doesn't change the name of the feature class—the dataset itself—which is still *City_ply*. A layer is a representation of the data, a rendering of the data, not the data itself. You can make any changes you want to a layer's properties without affecting the data that the layer is based on.

G Click OK on the Layer Properties dialog box to close it.

6) Get information about cities. Let's see what we can find out about the cities on the map.

A On the Tools toolbar, click the Identify tool ⓘ.

B Click any city polygon on the map.

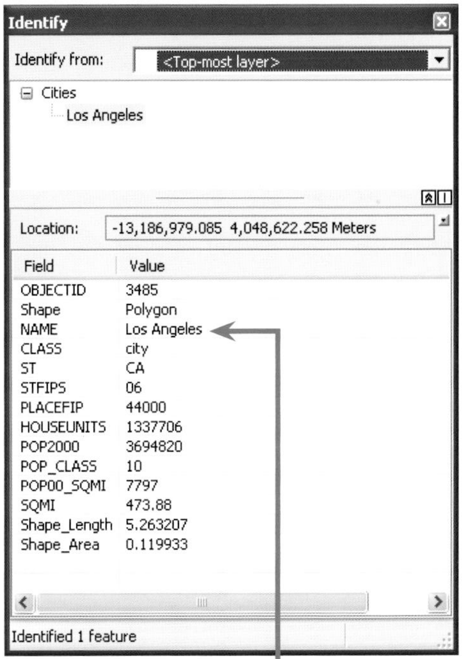

The city flashes green and an Identify window opens (Figure 1-12). In the top window, you see the name of the city you identified. In the bottom window, you see its attributes, or the information that this layer stores about cities. Some of them aren't too meaningful, but others are. If POP2000 is population for the year 2000, and POP00_SQMI is population per square mile for the same year, then you know that Los Angeles (if that's the city you identified) had 3,694,820 inhabitants at the time of the 2000 census and its population density was 7,797 people per square mile.

C Move the Identify window away from the map, if necessary. Click another city on the map.

The Identify window updates with information about the new city. All the cities have the same set of attributes; it's the values of the attributes that change.

D Identify a few more cities, then click somewhere on the basemap where there isn't a city.

The Identify window is empty and a message at the bottom says, "No identified features."

E Close the Identify window.

F In the Table of Contents, right-click *Cities* and choose Open Attribute Table (Figure 1-13).

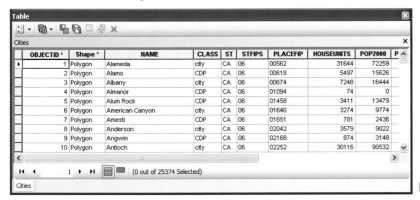

This example shows Los Angeles, but it doesn't matter which city you identify.

Figure 1-12

Figure 1-13

G Scroll across the table and look at the field names (the gray column headings).

This is a different presentation of the same information you saw when you identified cities.

H Scroll back to the beginning of the table, then scroll a little way down through the records (the table rows).

There are a lot of records in this table: in fact, 25,374 of them, as you can see at the bottom. Each corresponds to a unique feature—that is, a unique city—on the map. So there must be a lot of cities we're not seeing in the current view.

I Leave the table open, but move it out of the way of the map.

▷ You can move it outside the ArcMap application window.

J In the Table of Contents, right-click the *Cities* layer. On the context menu, choose Zoom To Layer.

The map zooms out to the geographic extent of the layer: the entire United States. You can't really distinguish individual cities at this scale.

K On the Tools toolbar, click the Go Back To Previous Extent button ◀.

7) **Select the record for Los Angeles.** When you select a record in an attribute table, the corresponding feature is selected on the map. (Likewise, when you select a feature on the map, its record is selected in the table.) Selections are marked with a blue highlight.

A Scroll up to the top of the table. Make sure the table is wide enough that you can see the POP2000 field (which contains population counts from the year 2000).

▷ If necessary, widen the table by dragging its edge.

B Right-click the the POP2000 field name. On the context menu, choose Sort Descending.

All 25,000 records are sorted in the order of their populations, from largest to smallest. Los Angeles is now the second record in the table, right after New York.

C At the left edge of the table, click the small gray box next to the Los Angeles record to select it (Figure 1-14).

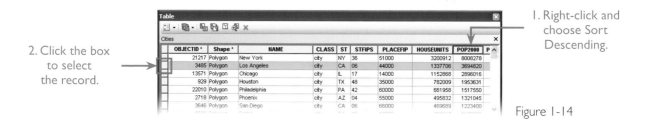

2. Click the box to select the record.

1. Right-click and choose Sort Descending.

Figure 1-14

On the map, the city of Los Angeles is highlighted in blue. You can probably see the whole city already, but let's make sure.

D Close the attribute table.

E In the Table of Contents, right-click *Cities* and choose Selection → Zoom to Selected Features.

The map zooms in close on the selected feature and centers it in the view (Figure 1-15). The city's odd shape is attributable to years of piecemeal expansion and incorporation. It has internal "holes" where it surrounds other cities, like Beverly Hills, or unincorporated areas. It also has a long, narrow southern corridor that connects it to its harbor at the Port of Los Angeles.

F On the Tools toolbar, click Clear Selected Features button ▣ to unselect the feature.

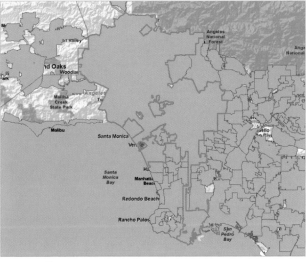

Figure 1-15

8) **Filter the display of cities with a definition query.** One of our project requirements is that the new park be inside the Los Angeles city limits. Therefore, we're naturally more interested in Los Angeles than in other cities. A layer property called a definition query lets us show only those features in a layer that interest us. A query, in general, is a logical expression that selects certain records in the attribute table. A definition query serves the specific purpose of showing only those records as features on the map and hiding all others.

A In the Table of Contents, right-click the *Cities* layer and choose Properties.

B In the Layer Properties dialog box, click the Definition Query tab (Figure 1-16).

C Click Query Builder to open the Query Builder dialog box.

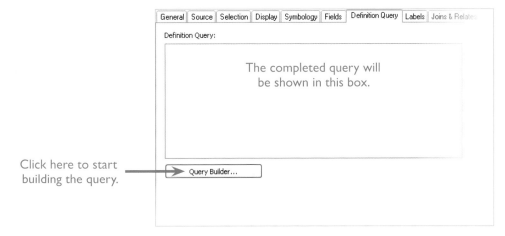

Figure 1-16

You build a query on an attribute table by specifying a field and setting a logical or arithmetic condition that values in that field have to satisfy. In our case, we want to find records with the value "Los Angeles" in the NAME field.

D In the list of field names at the top of the Query Builder dialog box, double-click "NAME."

The field is added to the expression box at the bottom of the Query Builder.

E Click the Equals button ⊡.

F Click Get Unique Values.

The middle box in the Query Builder is populated with a list of all the city names in the NAME field. Since there are so many, it would be inconvenient to scroll to the one we want.

G Click in the Go To box and start typing **Los Angeles**.

By the time you get to "Los An" the value "Los Angeles" will be highlighted in gray.

H Double-click Los Angeles.

I Compare your expression to Figure 1-17. If it doesn't match, click Clear and rebuild it.

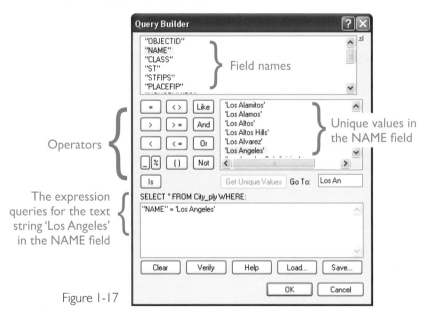

Figure 1-17

J Click Verify, then click OK on the Verifying Expression message box.

   ▷ If you get an error message, click OK. Clear and rebuild the expression.

K Click OK on the Query Builder dialog box.

The expression appears on the Definition Query tab (Figure 1-18).

◗ Click OK on the Layer Properties dialog box.

On the map, only the city of Los Angeles is shown. The other cities are hidden by the definition query (Figure 1-19).

Figure 1-18

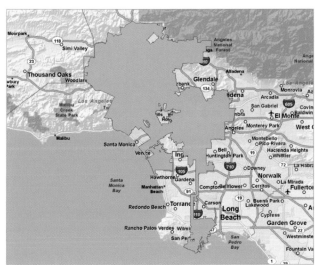

Figure 1-19

Ⓜ In the Table of Contents, right-click the *Cities* layer and choose Open Attribute Table.

The table shows just the record for Los Angeles.

Ⓝ Close the attribute table.

The other city features haven't been deleted. Clearing the definition query would display them again. Layer properties affect the display of data in a map, not the essential properties of the data itself: the number of features, their shapes, locations, and attributes.

Layers and datasets

A layer points or refers to a dataset stored somewhere on disk (as specified on the Source tab of the Layer Properties dialog box). A layer is *not* a physical copy of the data. The dataset stores the shapes, locations, and attributes of features; the layer stores display properties, including what the layer is named, which of its features are shown or selected, how those features are symbolized, and whether they are labeled. Changes to layer properties do not affect the dataset that the layer refers to. You can make as many layers as you want from the same dataset and give them different properties. These layers can coexist in the same map document or in many different map documents.

9) Add a layer of rivers. Let's add a layer of local rivers to the map and see where the Los Angeles River fits into the picture.

> **Ⓐ** On the Standard toolbar, click the Add Data button ⊞. (Click the icon itself, not the drop-down arrow.)

The Add Data dialog box opens to your last location. There's no river data here.

> **Ⓑ** In the row of buttons at the top of the Add Data dialog box, click the Up One Level button 🔼 (Figure 1-20).

Figure 1-20

> **Ⓒ** Double-click Hydro.

> **Ⓓ** Click *River* to select it, then click Add.

A layer of rivers is added to the map and placed at the top of the Table of Contents.

> **Ⓔ** On the Tools toolbar, click the Identify tool ⓘ, if necessary.

> **Ⓕ** Click on any river to identify it.

You see the name of the river (many of them don't have names) and its other attributes (Figure 1-21).

By default, the Identify tool identifies features from the topmost layer in the Table of Contents. If you miss a river, you'll either identify the City of Los Angeles or nothing. That's fine—leave the Identify window open and click again on a river.

> **Ⓖ** Click on a few more rivers to identify them. Try to identify the Los Angeles River.

The river runs west to east across the city, turns south near the city's eastern edge, and follows the 710 freeway to San Pedro Bay.

> **Ⓗ** Close the Identify window.

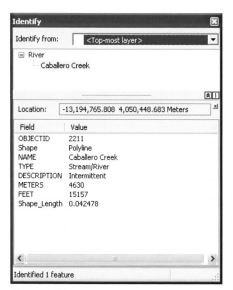

Figure 1-21

10) Make a definition query on the LA River. Just as we're mainly interested in one city, we're mainly interested in one river. We'll make another definition query to show just the Los Angeles River.

> **Ⓐ** In the Table of Contents, right-click the *River* layer and choose Properties.

> ▷ A shortcut is to double-click the layer name in the Table of Contents.

B Click the Definition Query tab, if necessary.

C Click Query Builder to open the Query Builder dialog box.

D In the list of field names at the top of the Query Builder dialog box, double-click "NAME."

E Click the Equals button = .

F Click Get Unique Values.

G Click in the Go To box and start typing **Los Angeles River**.

The value will be highlighted in gray before you finish typing.

H Double-click Los Angeles River.

I Compare your expression to Figure 1-22. If it doesn't match, clear Clear and rebuild it.

J Optionally, verify the expression.

K Click OK on the Query Builder dialog box.

L On the Layer Properties dialog box, click the Symbology tab.

M On the Symbology tab, click the button displaying the blue line symbol to open the Symbol Selector.

N Change the line width to 3 (Figure 1-23), then click OK on the Symbol Selector.

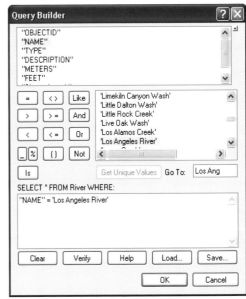

Figure 1-22

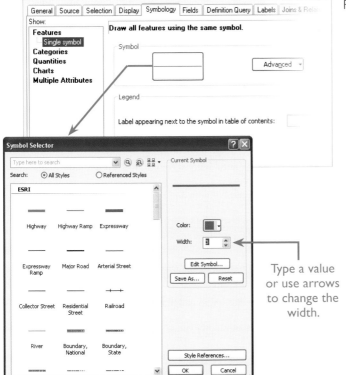

Figure 1-23

O On the Layer Properties dialog box, click the General tab.

P In the Layer Name box, delete the layer name *River* and type **Los Angeles River** (Figure 1-24).

> ▹ You can also rename a layer directly in the Table of Contents by clicking the name once to highlight it and then clicking it again. (Be careful not to double-click or you'll open the layer properties.)

Figure 1-24

Q Click OK on the Layer Properties dialog box.

On the map, the river is displayed with its new symbology (Figure 1-25). The Table of Contents reflects the new layer name.

11) Select Los Angeles River features with a query. On the map, the river looks like a single feature (just like the city), but it's not.

A In the Table of Contents, right-click the *Los Angeles River* layer and choose Open Attribute Table.

At the bottom of the table, you see that 0 of 17 records are selected. That means that, in this particular layer, the Los Angeles River is composed of 17 features. Why would that be?

B Scroll down through the table.

All the records have the same NAME value. Most have the same TYPE, but one is an artificial path. There are a few DESCRIPTION values. The need to maintain different attribute values for different parts of a geographic object is a common reason that data—especially data representing linear features like streets and rivers—is constructed this way (Figure 1-26). We'll come back to this point in the next lesson.

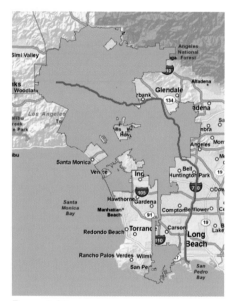

Figure 1-25

Figure 1-26

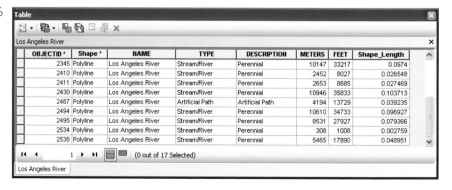

OBJECTID ▲	Shape ▲	NAME	TYPE	DESCRIPTION	METERS	FEET	Shape_Length
2345	Polyline	Los Angeles River	Stream/River	Perennial	10147	33217	0.0974
2410	Polyline	Los Angeles River	Stream/River	Perennial	2452	8027	0.026548
2411	Polyline	Los Angeles River	Stream/River	Perennial	2653	8685	0.027469
2430	Polyline	Los Angeles River	Stream/River	Perennial	10946	35833	0.103713
2487	Polyline	Los Angeles River	Artificial Path	Artificial Path	4194	13729	0.039235
2494	Polyline	Los Angeles River	Stream/River	Perennial	10610	34733	0.096927
2495	Polyline	Los Angeles River	Stream/River	Perennial	8531	27927	0.079366
2534	Polyline	Los Angeles River	Stream/River	Perennial	308	1008	0.002759
2538	Polyline	Los Angeles River	Stream/River	Perennial	5465	17890	0.048951

1 ▸ ▸▮ (0 out of 17 Selected)

Los Angeles River

C In the row of tools at the top of the table window, click the Select By Attributes button 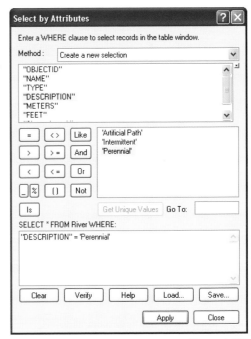.

Having noticed "perennial" and "intermittent" attribute values in the DESCRIPTION field, we might want to know which parts of the river are which. We can find out with an attribute query. An attribute query is like a definition query in that both single out features in a layer on the basis of attribute values. The difference is that an attribute query highlights (selects) features that satisfy the expression rather than hiding features that don't.

D At the top of the dialog box, make sure the Method drop-down list is set to "Create a New Selection."

E In the list of field names, double-click "DESCRIPTION" to add it to the expression box.

F Click the Equals button [=].

G Click Get Unique Values.

H In the list of unique values, double-click 'Perennial.'

I Confirm that your expression matches the one in Figure 1-27, then click Apply.

Figure 1-27

Twelve records are selected in the table. The corresponding features are selected on the map, showing you that the river is perennial for most of its length (Figure 1-28).

It's not among our guidelines to locate the new park along the river's perennial stretch, but it's interesting that fairly simple data exploration may introduce new ways of thinking about your problem.

J Close the Select By Attributes dialog box.

K At the top of the table, click the Clear Selection button.

Clearing the record selection clears the feature selection on the map. (The reverse is true as well.)

L Close the attribute table.

Selecting features

You make selections in order to work with a subset of features in a layer. You can use a selection to zoom the map in to a specific area, to make a new layer from the selected subset, to get statistical information about the subset, or for many other things.

Selections are also used in queries. Whether you do an attribute query (to find records with a certain attribute value) or a spatial query (to find features with a certain spatial relationship to other features), the records and features that satisfy the query are returned as a selection on the layer.

Figure 1-28

12) Save the map document. This is a good time to save the map document. Save your maps often as you do these exercises—don't wait to be reminded.

Ⓐ On the Standard toolbar click the Save button 💾.

In the Save As dialog box, ArcMap should default to your My Documents\ArcGIS folder. In this book, we'll save most of our maps to the ParkSite\MapsAndMore folder.

Ⓑ In the Save As dialog box, click the Up One Level button 🔼 twice (or more if necessary) to navigate to your Desktop.

Ⓒ Double-click My Computer and navigate to C:\UGIS\ParkSite\MapsAndMore.

Ⓓ Replace the default file name with **Lesson1** (Figure 1-29), then click Save.

ArcMap will add the .mxd file extension automatically.

When you save a map document, ArcMap defaults to the location where your last map document was saved. With no save history, it defaults to MyDocuments\ArcGIS.

Figure 1-29

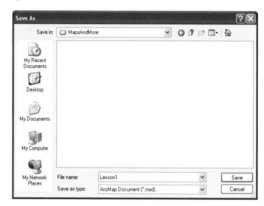

13) Add a basemap of imagery. Imagery provides a detailed, photorealistic view of the ground, and we'll rely on it to explore the LA River in more detail. Imagery also has other important uses, like providing a background against which to edit features (Lesson 5), and "ground-truthing" analysis results (Lesson 6).

Ⓐ On the Standard toolbar, click the drop-down arrow next to the Add Data button and choose Add Basemap.

Ⓑ In the Add Basemap window, click *Imagery* to select it (Figure 1-30), then click Add.

The selection and arrangement of basemaps is subject to change, so your screen may look different.

Figure 1-30

The imagery basemap is added as a layer, covering the streets basemap. ArcMap applies some logic to the layer order in the Table of Contents. The new basemap goes above the old one but underneath the layers that aren't basemaps.

Now there are two layers called *Basemap* in the Table of Contents. ArcMap doesn't mind, but it might be a little confusing for us.

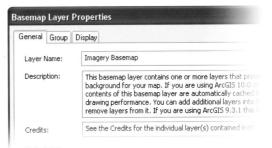

C In the Table of Contents, open the layer properties for the *Basemap* imagery layer you just added.

D In the Basemap Layer Properties dialog box, click the General tab, if necessary. Rename the layer **Imagery Basemap** (Figure 1-31), then click OK.

E Rename the other basemap layer *Streets Basemap* and click OK on its Layer Properties dialog box.

F In the Table of Contents, uncheck the *Streets Basemap* layer to turn it off.

▷ You can leave its *Streets* sublayer turned on.

G Under the *Imagery Basemap* layer, turn on the *Boundaries and Places* and *Transportation* sublayers.

Depending on your Internet connection speed, the redraw time may vary. (If the small globe 🌐 is spinning in the lower part of the ArcMap application window, it means the layer is still loading.) Your map and Table of Contents should look like (Figure 1-32).

Figure 1-31

Basemap layers have fewer properties to set than layers based on your own data.

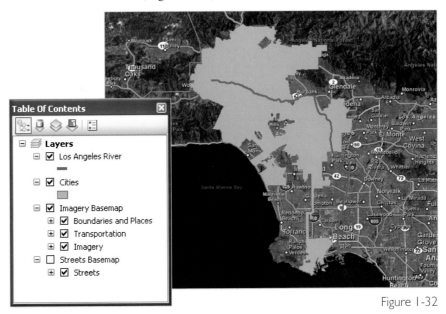

Here the Table of Contents is shown floating free. By default it's docked in the ArcMap application window, which is probably where it is on your screen.

Figure 1-32

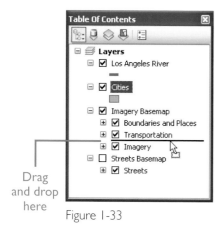

Figure 1-33

Drag
and drop
here

14) Make a layer sandwich. The *Imagery Basemap* layer is a group layer consisting of three sublayers. Right now, the *Cities* layer draws on top of the group layer, covering everything underneath Los Angeles. We can insert the *Cities* layer into the group layer to make a kind of sandwich.

Ⓐ In the Table of Contents, click and drag the *Cities* layer slowly downward.

As you drag, the layer's position in the Table of Contents is represented by a horizontal black bar.

Ⓑ When the black bar is positioned between the *Transportation* and *Imagery* sublayers (Figure 1-33), release the mouse button.

In the Table of Contents, the *Cities* layer appears where you placed it. On the map, the city of Los Angeles now draws above the imagery, but underneath the place names and roads (Figure 1-34).

If yellow warning signs appear next to the *Imagery Basemap* layer or its sublayers, you can ignore them.

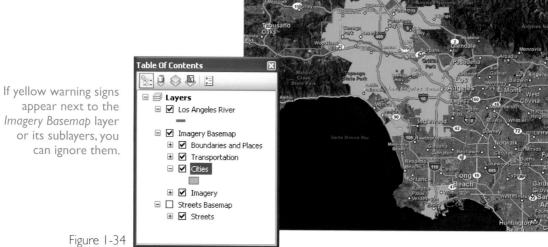

Figure 1-34

Group layers

Most combinations of layers can be grouped or ungrouped in a map. If you have a lot of layers in your map, grouping them can help keep the Table of Contents organized. Group layers can be treated as a unit with respect to properties like visibility, transparency, and position in the Table of Contents—without your giving up the ability to set these properties independently for each sublayer. To make a group layer, select the layers you want in the Table of Contents by holding the Control key and clicking each layer. Then right-click any selected layer and choose Group. To ungroup a layer, right-click it in the Table of Contents and choose Ungroup.

15) Create a spatial bookmark. Certain views in a map are useful for orientation or reference. You can bookmark a view to make it easy to return to.

Figure 1-35

ⓐ From the main menu, choose Bookmarks → Create.

ⓑ In the Spatial Bookmark dialog box, replace the default name with **City of Los Angeles** (Figure 1-35). Click OK.

ⓒ Use the Zoom In tool ⊕ to zoom in on the river's mouth in the harbor at Long Beach.

ⓓ On the Standard toolbar, click the Map Scale drop-down arrow and choose 1:10,000 (Figure 1-36).

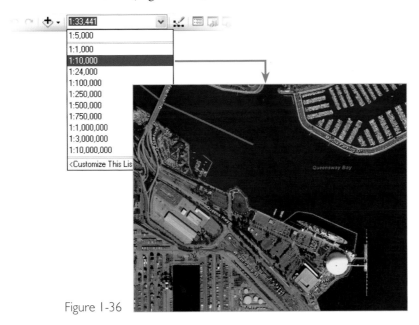

At this scale, one unit of measure on the map is equivalent to ten thousand of the same units on the ground. Loosely, a thing on the map is ten thousand times smaller than its actual size.

Figure 1-36

ⓔ Use the zoom and pan tools to explore the harbor.

The imagery is very high resolution and you'll see more detail as you keep zooming in. Eventually, you'll reach a limit.

ⓕ When you're ready, from the main menu, choose Bookmarks → City of Los Angeles.

The map view returns to the bookmarked extent.

16) Change the symbology for Los Angeles. The boundary of Los Angeles is filled in with a solid color, so the imagery underneath is still covered up. Pretty soon we're going to zoom in and follow the river's course through the city. It will serve that purpose to resymbolize the city so that we just see its outline.

Ⓐ In the Table of Contents, click and drag the *Cities* layer out of the *Imagery Basemap* layer. Drop it at the top of the Table of Contents.

Ⓑ Open the layer properties for the *Cities* layer. Click the Symbology tab, if necessary.

Ⓒ On the Symbology tab, click the button displaying the current color symbol to open the Symbol Selector.

Ⓓ In the Symbol Selector, click the Fill Color button to open the color palette.

▷ You can click either the color square itself or the drop-down arrow.

Ⓔ At the top of the color palette, click No Color.

This will make the symbol transparent.

Ⓕ Change the Outline Width value to 2.

Ⓖ Click the Outline Color button. On the color palette, click Autunite Yellow (Figure 1-37).

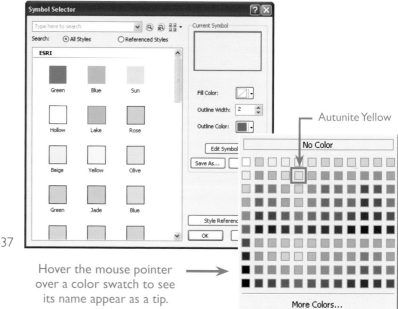

Figure 1-37

Hover the mouse pointer over a color swatch to see its name appear as a tip.

O Click OK on the Symbol Selector. Click Apply on the Layer Properties dialog box and leave it open.

The new symbology is displayed on the map (Figure 1-38).

I Click the General tab. Rename the layer **Los Angeles**.

J Click OK on the Layer Properties dialog box.

17) Save the layer as a layer file. It wasn't hard to make this particular symbol, but often it does take time and effort to create good symbology. Having done so, you may want to reuse that symbology. In this book, for instance, we'll probably want to draw the outline of Los Angeles in other map documents in coming lessons.

You can save layer properties to a file called a layer file, which has the file extension *.lyr*. A layer file is not a copy of the data, but you add it to a map document in the same way that you add data. The layer file stores all the properties of a layer—name, symbology, definition query, and so on—including the path to the layer's source dataset. When you add a layer file to ArcMap, the layer draws with its properties already set.

A In the Table of Contents, right-click *Los Angeles* and choose Save As Layer File (near the bottom of the context menu).

The Save Layer dialog box opens to the ParkSite\MapsAndMore folder (Figure 1-39).

B Accept the default name *Los Angeles.lyr* and click Save.

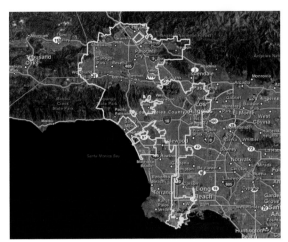

Figure 1-38

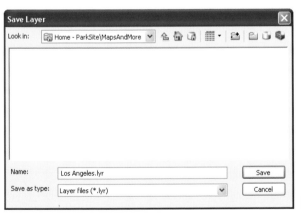

When you saved the map document in the MapsAndMore folder, this became the map's "home" folder. Now when you save other files associated with the map, like layer files or bookmarks, they will be saved to this location by default.

Figure 1-39

We'll remove the *Los Angeles* layer that's currently in the map, then add the layer file to see how it works.

C In the Table of Contents, right-click *Los Angeles* and choose Remove.

The layer disappears from the map and the Table of Contents.

D On the Standard toolbar, click the Add Data button.

E In the Add Data dialog box, navigate up to the ParkSite folder, then double-click the MapsAndMore folder.

F Click *Los Angeles.lyr* to select it (Figure 1-40), then click Add.

Figure 1-40

The layer is added to the map with its properties set. Its position in the Table of Contents has changed: by default, ArcMap places polygon layers above basemap layers but below layers of point or line features.

G Open the layer properties for *Los Angeles*. In the Layer Properties dialog box, click the Source tab.

As before, the layer points to the *City_ply* feature class.

H Close the Layer Properties dialog box.

Any time you add the *Los Angeles.lyr* file to a map document, the layer will draw as a yellow outline with a definition query on the city of Los Angeles. Once the layer has been added to the map, however, it's just the same as any other layer, and you can change its properties however you want. (Not that we want to change them.)

18) Follow the river. Let's start developing a sense of the study area by following the river's course through the city.

A Maximize the ArcMap window if you haven't done so already.

B Use the Zoom In tool to zoom in on the river's "source" in the community of Canoga Park in northwest Los Angeles.

The river officially starts where Bell Creek and the Arroyo Calabasas converge at Canoga Park High School. (You can see these two tributaries on the *Imagery Basemap* layer. You don't see them as features because they've been excluded by the definition query.)

C On the Standard toolbar, click the Map Scale drop-down arrow and choose 1:24,000.

D Click the Pan tool and slowly pan east along the river.

Densely populated residential neighborhoods line both sides of the river until you get to the Sepulveda Basin (Figure 1-41), a large recreational area with golf courses and a lake. The river bottom is natural here, becoming concrete again at the Sepulveda Dam in the southeastern corner of the basin.

East of the Sepulveda Basin, the river follows the Hollywood Freeway (the 101) for a while, and is again surrounded by fairly dense residential and commercial areas.

E On the keyboard, press and hold the Q key.

Now you roam continuously across the display in whichever direction you point the mouse. To control your speed, make small brushing movements with the mouse either with or against the grain of your movement. As you roam, basemap layers should draw smoothly and continuously, although your experience may vary. Other layers, like the *Los Angeles River* layer, suspend drawing and catch up when you stop.

F Release the Q key to stop roaming.

▷ You can also use the four arrow keys on the keyboard to roam.

G Continue to pan (or roam) along the river.

After crossing the 101 a second time, the river flows along the northern edge of Universal Studios (Figure 1-42). It continues east, then bends sharply south as it curves around Griffith Park (at 4,218 acres, one of the largest city parks in the United States). The city boundary zigzags back and forth across the river here. To the north of Griffith Park lies the city of Burbank; to the east is Glendale.

As it flows south, the river runs parallel to the Golden State Freeway (the 5). It dips away, then comes back to cross both the 5 Freeway and the Harbor Freeway (110) at Elysian Park. Elysian Park is the home of Dodger Stadium, where the Los Angeles Dodgers play major league baseball.

Figure 1-41

Sepulveda Basin Recreation Area

Figure 1-42

Universal Studios

19) Save bookmarks as place files. Dodger Stadium is a landmark that we may want to return to.

Ⓐ Center your view on Dodger Stadium, more or less as shown in Figure 1-43.

Ⓑ From the main menu, choose Bookmarks ➤ Create.

Ⓒ In the Spatial Bookmark dialog box, name the bookmark **Dodger Stadium** (Figure 1-44), and click OK.

Figure 1-43

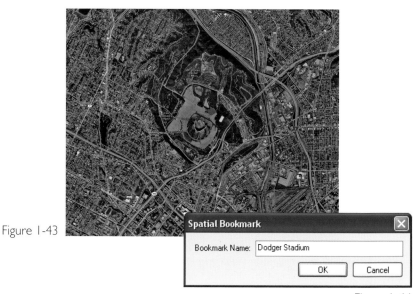

Figure 1-44

Bookmarks are meant for use within a particular map document. The two we've created so far, however, might be useful in other maps. There's a tool for managing places so that any map document can access them: it's called My Places and we'll use it in Lesson 3. For now, we'll save our bookmarks as external files called "place files." When the time comes, we'll be able to load them into My Places.

Ⓓ From the main menu, choose Bookmarks ➤ Manage.

You should see two bookmarks appear in the Bookmarks Manager, with City of Los Angeles highlighted on top (Figure 1-45).

Ⓔ Near the bottom of the Bookmarks Manager dialog box, click Save ➤ Save All.

Figure 1-45

The Save Bookmarks dialog box opens to the MapsAndMore folder.

- **F** In the File name box, type **Lesson1 Places** (Figure 1-46). The file type is an ArcGIS Place File with the extension .dat. Click Save.

- **G** Close the Bookmarks Manager dialog box.

20) Pan to the city limits. We'll follow the river until it crosses the LA city limits, which marks the boundary of our study area.

- **A** Pan along the river as it runs south.

This last section of river passes through an industrial landscape. You'll cross the 101 Freeway (for the third time), then the Santa Monica Freeway (the 10) with downtown Los Angeles to your west. The river leaves the city amidst a tangle of railroad tracks at Redondo Junction.

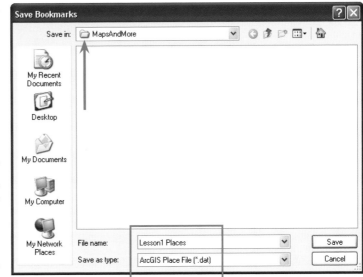

Figure 1-46

- **B** Save the map.

- **C** If you're going on to the next exercise now, leave ArcMap and Lesson1.mxd open. Otherwise, from the main menu, choose File → Exit.

Exercise 1b: Do exploratory data analysis

In this exercise, we'll add park data and census data (containing demographic and socioeconomic information) to our map. Our goal is to pay attention to patterns in the data and thereby build an intuitive sense of likely and unlikely locations for the park. This intuition should give us confidence that the analysis results we get in Lesson 6 are plausible. Conversely, if the results contradict our gut feeling, we may be alerted to possible mistakes in the analysis.

1) Start ArcMap. We'll start ArcMap, if necessary, and continue working with our map document from the last exercise.

- **A** If ArcMap is open from the last exercise, go to Step 2.

- **B** Otherwise, start ArcMap.

  ▷ Use the Start menu or double-click the ArcMap 10 shortcut on your computer desktop.

In the ArcMap Getting Started dialog box, you should see Lesson1 under the heading "Recent."

C Click Lesson1 to highlight it (Figure 1-47) and click Open.

Figure 1-47

The map opens to a view of where the LA River leaves Los Angeles and enters the industrial city of Vernon.

2) **Add a layer of parks.** One of our geographic constraints is that the new park not be located too near existing parks. We know there are a couple of big parks along the river, but let's look at the whole situation.

A On the right side of the ArcMap application window, click on the Catalog tab to open the Catalog window (Figure 1-48).

▷ If you don't have this tab, on the Standard toolbar, click the Catalog window button 🗺.

The Catalog window lets you view and manage geographic files on disk. The top entry in the window is your home folder, which presently contains your map document, the layer file you saved in the last exercise, and some graphs that are part of the starting data and which we'll use in Lesson 8. By default, the catalog recognizes only a limited number of file types to help you focus on geographic data.

Click the Auto Hide button to toggle the window's state

Figure 1-48

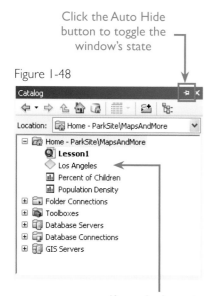

If you don't see Los Angeles.lyr, right-click the Home folder and choose Refresh.

The Catalog window has two states. When the pin icon in the upper right corner points sideways, the window is "shy" and closes if not in use. When the pin icon points down, the window stays open. In this state, you can move and resize the window. The shy state is nice because it uses screen space efficiently, but you should work in the way you like best.

B In the Catalog window, click the plus sign next to Folder Connections to expand it.

C Expand C:\UGIS.

This is the folder connection you created in the last exercise from the Add Data dialog box.

D Expand the following items:

- ParkSite
- SourceData
- ESRI
- Landmark

The Catalog window should look like Figure 1-49.

E In the Catalog window, click *Parkland* to select it.

F Drag and drop *Parkland* anywhere over the ArcMap window.

The layer is added to the Table of Contents and the map. Dragging and dropping data from the Catalog window is the same as adding it with the Add Data button—it's just faster.

3) Symbolize the layer. A shade of green is usually the right cartographic choice for parks. ArcMap may have chosen one by a stroke of luck, but probably not. We'll symbolize the layer after taking a look at its extent.

A In the Table of Contents, right-click the *Parkland* layer and choose Zoom To Layer (Figure 1-50).

Figure 1-50

The data extends well beyond Los Angeles.

B In the Table of Contents, turn *Parkland* off and on a few times by clicking its check box.

C When you've seen the imagery underneath, leave the layer turned on.

The basemap shows that those big polygons to the north correspond to mountains. They're probably national forests.

Figure 1-49

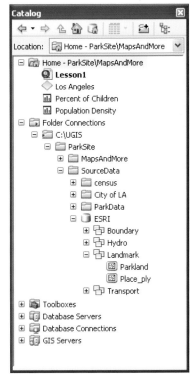

In figures, the Catalog window is usually shown floating free from ArcMap (as here).

Figure 1-51

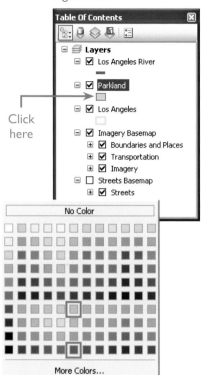

Figure 1-52

D In the Table of Contents, click the color patch under the *Parkland* layer name to open the Symbol Selector (Figure 1-51).

▷ This is a shortcut to opening the Symbol Selector from the layer properties.

E In the Symbol Selector, click the Fill Color button to open the color palette. Click Apple Dust (Figure 1-52).

F Click the Outline Color button to open the color palette again. Click Moss Green (Figure 1-52).

G Click OK on the Symbol Selector.

The layer symbology is updated on the map.

4) Identify features. Let's find out what attributes this layer has.

A On the Tools toolbar, click the Identify tool 🛈 .

B Click on one of the big park features in the mountains to identify it.

Among the attributes are the park's name, its type, and its acreage (Figure 1-53).

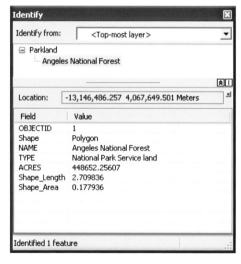

Figure 1-53

C Leave the Identify window open. From the main menu, choose Bookmarks → City of Los Angeles.

At this scale, you can make out the larger parks within the city.

D Try to identify the parks we noticed on page 29: the Sepulveda Basin Recreation Area, Griffith Park, and Elysian Park.

E When you're done, leave the Identify window open and zoom to the Dodger Stadium bookmark.

❻ Identify some of the neighborhood-sized parks in the view, like Alpine Park and Downey Playground.

Our list of park requirements didn't have an upper size limit, but we can expect our candidates to be under ten acres. Our trip down the river in the last exercise didn't reveal any great big tracts of open land that weren't already parks.

❼ Close the Identify window.

5) Label parks. Labeling the parks will let us see their names at a glance.

Ⓐ In the Table of Contents, right-click *Parkland* and choose Label Features.

The parks are labeled with their names (Figure 1-54), which come from the NAME attribute in the *Parkland* attribute table.

The default label symbology isn't ideal for this map. For one thing, the black text disappears into the imagery wherever the label doesn't fit inside the park.

Ⓑ Open the layer properties for the *Parkland* layer and click the Labels tab (Figure 1-55).

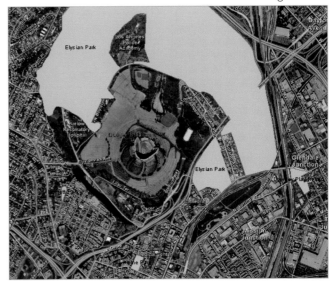

Figure 1-54

ArcMap places labels dynamically. The position of labels in a view depends on the map scale, the extent, the presence of other features, and more. The label placement in your map may be different from what's shown here.

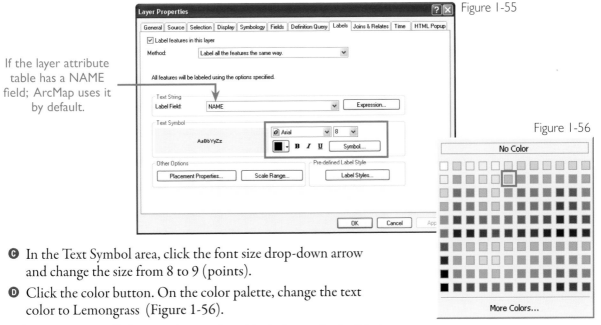

Figure 1-55

If the layer attribute table has a NAME field; ArcMap uses it by default.

Figure 1-56

Ⓒ In the Text Symbol area, click the font size drop-down arrow and change the size from 8 to 9 (points).

Ⓓ Click the color button. On the color palette, change the text color to Lemongrass (Figure 1-56).

Light green text will show up better than black, but to be legible it will still need an outline, or halo.

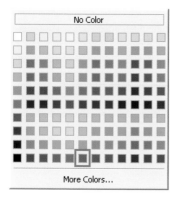

Figure 1-57

E Click the Symbol button to open the Symbol Selector.

F On the Symbol Selector, underneath the properties of the current symbol, click Edit Symbol.

G In the Editor dialog box, click the Mask tab.

H Click the Halo option and change the size to 1.

The default halo color is white, but we're going to use something darker.

I Click Symbol.

The Symbol Selector opens again.

J Click the Fill Color button and change the fill color (the halo color) to Moss Green (Figure 1-57).

K Click OK on the Symbol Selector.

Your Editor dialog box should look like Figure 1-58.

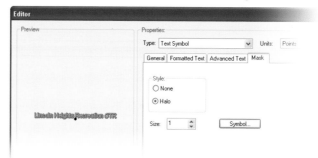

Figure 1-58

L Click OK on all open dialog boxes.

On the map, the park labels are visible against the basemap (Figure 1-59).

Figure 1-59

6) Set a scale range for the labels. The park labels look good at this fairly large scale, where there's room to accommodate them. Let's see what happens when we zoom out.

Ⓐ Zoom to the City of Los Angeles bookmark.

At this smaller (zoomed-out) scale, the labels overwhelm the map (Figure 1-60). We could just turn them off, of course, whenever they seem to be too crowded. A nicer solution is to make their visibility dependent on the map scale.

Ⓑ Open the layer properties for the *Parkland* layer. Click the Labels tab if necessary.

Ⓒ Near the bottom of the Labels tab, click Scale Range.

Ⓓ In the Scale Range dialog box, click the "Don't show labels when zoomed" option.

Ⓔ Click in the "Out beyond" box and type **40,000**.

Whenever the map scale crosses the 1:40,000 threshold, the park labels will turn off automatically.

Ⓕ Compare your Scale Range dialog box to Figure 1-61, then click OK. Click OK again on the Layer Properties dialog box.

The park labels disappear from the map.

Ⓖ Zoom to the Dodger Stadium bookmark.

As long as the map scale is larger than 1:40,000 (which it should be), the labels show up again. We could make further improvements to their appearance, but this is enough for our present needs. One obvious enhancement would be to break the longer park names into multiple lines. In Lesson 8, we'll see that ArcMap has a sophisticated label generator called Maplex™ that does this automatically.

Figure 1-60

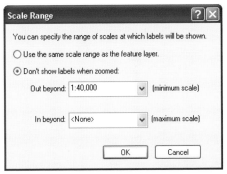

Figure 1-61

Scale ranges for layers

You can set scale ranges for layers as well as for labels. When you design maps to be viewed at multiple scales, you normally want layers representing detailed data, like buildings or utility lines, to be visible only at large (close-up) scales. You may want your map to include multiple copies of a layer, each with a different scale range and unique symbology. For example, one layer might represent trees with an unobtrusive generic point symbol at medium scales. A second layer might represent the same trees with a more detailed and realistic symbol at large scales. Scale ranges for layers are set on the General tab of the Layer Properties dialog box.

7) Dim the basemap. Without doing anything further to the labels, we can emphasize them a little more by dimming the basemap. To do that, we have to add a toolbar to the interface.

Ⓐ From the main menu, choose Customize → Toolbars → Effects.

The Effects toolbar is added (Figure 1-62). ArcMap has a lot of special-purpose toolbars, and we'll use several of them in this book.

Figure 1-62

| Effects ▾ ✕ |
| Layer: ✧ Los Angeles River ▾ ◑ ◆ ◇ ⬓ ◇ 500 ⬍ |

Ⓑ Drag the toolbar to a position you like.

▷ You can dock it on the top, bottom, or either side of the ArcMap window, or leave it undocked anywhere on your screen.

Ⓒ On the Effects toolbar, click the Layer drop-down arrow and choose *Imagery Basemap*.

Ⓓ Click the Adjust Dim Level button ◇ to open a slider bar.

Ⓔ Set the Dim Level to 10% (Figure 1-63).

On the map, the imagery fades slightly.

Dim Level

10 %

Figure 1-63

8) Add a layer of census tracts. The U.S. Census Bureau gathers socioeconomic data about households and organizes it by various units of geographic generality. One of these units is a census tract, which is a relatively small subdivision of a county.

Ⓐ Open the Catalog Window.

Ⓑ Under the SourceData folder, expand the census folder.

Ⓒ Drag and drop *tracts* onto the map.

In the Table of Contents, the *tracts* layer is added above *Parkland*. On the map, the tracts cover the parks (but not the park labels).

Ⓓ In the Table of Contents, right-click the *tracts* layer and choose Zoom To Layer.

The census tract data covers Los Angeles County, including the islands of Catalina and San Clemente.

Ⓔ Zoom to the City of Los Angeles bookmark.

The *tracts* layer is made up of contiguous polygons that are reminiscent of a jigsaw puzzle (Figure 1-64). Since each piece of the puzzle represents a different set of living, breathing human beings, it's natural that each tract's attribute values for population, income, age, and so on, are different.

Figure 1-64

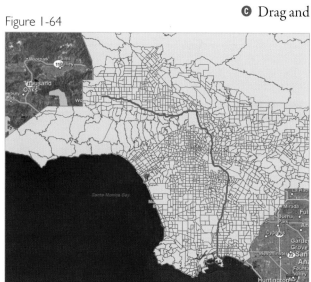

9) **Open the attribute table.** Let's see what attributes the table has.

Ⓐ In the Table of Contents, right-click the *tracts* layer and choose Open Attribute Table.

Ⓑ Scroll across the table and look at the field names.

The table has some identification codes, followed by population attributes and breakdowns of ethnicity, sex, age, household status, and housing. Some of the field names are fairly easy to interpret, others less so. In Lesson 2, we'll see how to get information about our information (metadata).

Ⓒ Locate the POP07_SQMI field (Figure 1-65).

FIPS	POP2000	POP2007	POP00_SQMI	POP07_SQMI	WHITE	BLACK	AMERI_ES	ASIAN	HAWN_PI	OTH
06037101110	4500	4801	10227.↓	10911.4	3196	86	41	329	4	
06037101120	3280	3640	658.↓	730.9	2670	60	17	195	4	
06037101210	6066	6303	2426↓	25212	3427	344	63	341	1	1
06037101220	3028	3269	11214.↓	12107.4	2018	58	32	162	1	
06037101300	3974	4080	4096.↓	4206.2	3371	22	23	261	2	
06037101400	3760	3862	1613.↓	1657.5	2985	42	25	238	4	
06037102101	3387	3452	45↓	462.7	2609	69	37	183	2	
06037102102	6739	7067	4522.↓	4743	4683	141	38	681	4	

(0 out of 2054 Selected)

Figure 1-65

This attribute stores population density (people per square mile) for the year 2007. Since one of our criteria is to locate the park in a densely populated area, this is relevant information. The attribute doesn't tell us what value should be a threshold for "high density," but it does give us a way to start making patterns on a map.

Ⓓ Right-click the POP07_SQMI field name and choose Statistics (Figure 1-66).

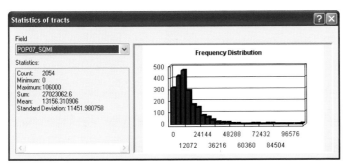

Figure 1-66

The Statistics of tracts window displays summary statistics for the field. The lowest value is 0 (at least one tract must be unpopulated) and the highest is 106,000. The frequency distribution chart, or histogram, on the right shows you that most of the values are between 0 and 25,000. The remaining values spread out in a long tail.

❺ Close the Statistics of tracts window.

❻ Close the table.

10) Symbolize census tracts by population density. Symbolizing a layer by an attribute, also called thematic mapping, allows us to see how values are spatially distributed.

🅐 Open the layer properties for the *tracts* layer and click the Symbology tab.

By default, all features in a layer have a single symbol (Figure 1-67). That's why all your census tracts are light blue, or whatever color they happen to be.

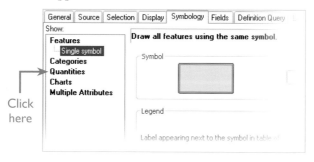

Click here

Figure 1-67

🅑 In the Show box, click Quantities.

Under Quantities, four methods for symbolizing numeric attributes are listed. The default Graduated colors method will give each feature in the *tracts* layer a color that reflects its numeric value in a specific field.

🅒 In the Fields area, click the Value drop-down list and choose POP07_SQMI (Figure 1-68).

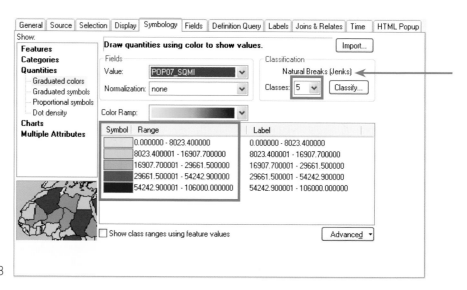

Figure 1-68

A lot is going on here. The values in the field, ranging from 0 to 106,000, are divided into five classes. The starting and ending values for each class are calculated by a "natural breaks" algorithm that separates clumps in the data. That's why the range of values is different from class to class and why classes break at eccentric numbers. Each class is associated with a symbol in a color ramp (probably yellow to dark red, but yours may be different).

D Click Apply. Move the Layer Properties dialog box out of the way.

On the map, the tracts are symbolized by population density (Figure 1-69).

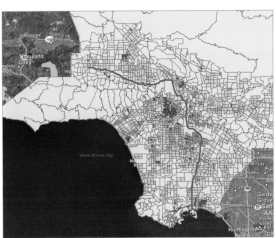

Figure 1-69

11) Change the classification. Quantitative symbology is flexible, and you can present data in many ways. Because all we want is a general sense of viable areas for our project, and because we're going to look at a couple of variables together, we should keep our presentation simple.

A In the Layer Properties dialog box, click the Classes drop-down arrow and choose 3.

B Click Classify to open the Classification dialog box (Figure 1-70).

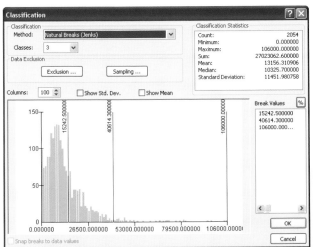

Figure 1-70

The histogram shows you the distribution of values in relation to the current class breaks. We can change the algorithm used to set class breaks, or make manual adjustments as desired.

C Click the Method drop-down arrow and choose Equal Interval (Figure 1-71).

D Now set the Method to Quantile.

Figure 1-71

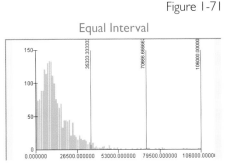

Equal Interval

Classes are evenly spaced. In this case, almost all the records fall in the first class.

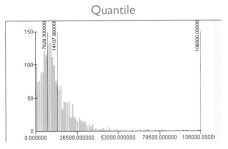

Quantile

Each class has an equal number of records. Class spacing is very different.

With three classes, we'll easily be able to see high, medium, and low values. The Quantile method guarantees that an equal number of tracts will fall into each class. It should be understood that there are no inherently good or correct ways to classify data—there are only ways that serve the purpose of your map.

E In the Break Values box to the right of the histogram, click the the first class break point (7629.3) to select it.

The value becomes editable. At the bottom of the dialog box, a message tells you that there are 684 elements (that is, records) in this class.

F Replace the highlighted value with **8,000** and press Enter.

The second class break point is selected and editable.

G Replace the highlighted value (14107.9) with **16,000**.

H Click in some white space in the Histogram window to stop editing class breaks (Figure 1-72).

At the top of the dialog box, the classification method has been reset to Manual because we've changed the class breaks. The histogram is updated, too. We no longer have a pure quantile classification, but having our classes break at round numbers makes intuitive sense.

I Click OK on the Classification dialog box.

12) Change the symbology. We'll make some changes to the symbology as well.

A On the Layer Properties dialog box, click the Color Ramp drop-down arrow (or click on the ramp itself).

B Scroll up and down to see the choices, then click the ramp again to close the list.

C Right-click on the color ramp. On the context menu, click Graphic View to uncheck it (Figure 1-73).

Figure 1-72

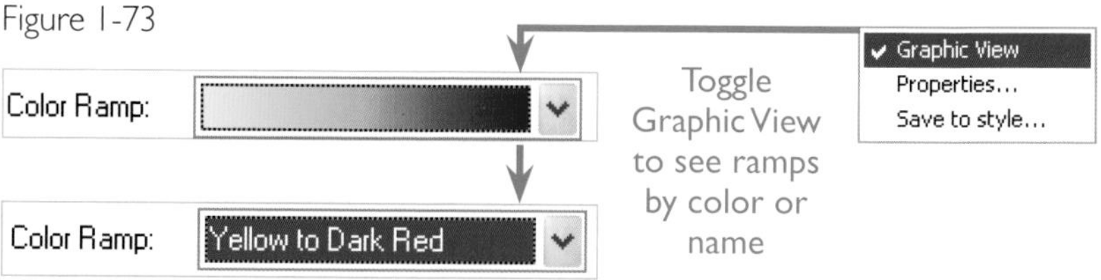

Figure 1-74

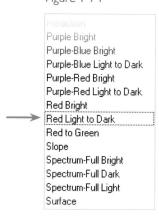

Figure 1-73

D Click on the ramp's name. In the drop-down list, click the Red Light to Dark ramp to select it (Figure 1-74).

E Right-click the ramp name and choose Graphic View to show the ramps by color again.

Underneath the color ramp is a box with three columns:

- The Symbol column shows the symbol for each class.
- The Range column shows the range of values for each class.
- The Label column shows how each symbol will be described in the Table of Contents. (By default, the label matches the range.)

F Click the Symbol column heading and choose Properties for All Symbols.

G In the Symbol Selector, click the Outline Color button. On the color palette, choose No Color.

H Click OK on the Symbol Selector.

We're taking away the outlines because we don't need to see the tract boundaries on the map. For now, we're interested in them as areas, not specifically as tracts.

I In the Label column, click on the first label (0.000000 - 8000.000000) to make it editable. Type **Low** and press Enter.

J Replace the second label with **Medium**. Press Enter.

K Change the third label to **High**. Click outside the edit box to commit the edit.

L Compare your settings to (Figure 1-75), then click Apply.

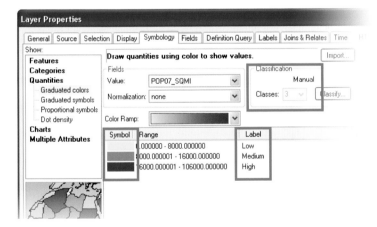

Figure 1-75

M In the Layer Properties dialog box, click the Display tab. In the Transparent % box at the top, replace the value 0 with **50** (Figure 1-76).

Figure 1-76

N Click OK on the Layer Properties dialog box.

O In the Table of Contents, drag and drop the *tracts* layer underneath the *Los Angeles* layer.

On the map, we can now see where population is concentrated along the river, and we can see it in relation to existing parks (Figure 1-77).

Figure 1-77

13) Measure distance from the river to parks. Making a few measurements will improve our ability to estimate distance on the map and will give us a better intuitive sense of how close to the river the new park should be.

A Zoom to the Dodger Stadium bookmark.

B Pan the map so that a number of parks are in the view. Feel free to zoom in or out.

C On the Tools toolbar, click the Measure tool ⬌ to open the Measure dialog box.

D In the row of tools at the top of the Measure dialog box, make sure the Measure Line tool is selected (Figure 1-78).

E Click the Choose Units button, then choose Distance → Miles.

Confirm that Measure Line is selected.

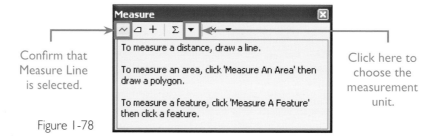

Click here to choose the measurement unit.

Figure 1-78

On the map, the mouse pointer icon changes to a ruler with an inscribed crosshairs.

❻ Move the mouse pointer over a park, such as Cypress Park (northeast of Dodger Stadium on the east side of the river).

The mouse pointer "snaps" to the park boundary. You can probably notice notice this effect in the way the mouse moves. It's confirmed on-screen by a light gray square near the mouse pointer and a message like "Parkland: Vertex" or "Parkland: Edge" (Figure 1-79).

❼ Click to start a measurement.

❽ Move the mouse pointer (you don't have to drag) to the river.

Figure 1-79

The mouse pointer snaps to the river feature. The measurement result is displayed in the Measure dialog box (Figure 1-80).

❾ Double-click to end the measurement.

❿ Click on another park and measure its distance to the river.

The new result replaces the previous one in the Measure dialog box.

Figure 1-80

⓫ Measure the distances from a few more parks.

Cypress Park, Elysian Valley Rec Center Park, and Downey Playground are close to the river. Elyria Canyon Park, a little over three quarters of a mile away (at its nearest edge), stretches the notion of proximity. Bear in mind that these measurements are straight-line distances, not distances along streets that people would take from their house to a park or from a park to the river.

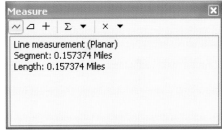

Your result will depend on where you start and end your measurement.

⓵ Close the Measure dialog box.

⓶ Zoom to the City of Los Angeles bookmark.

14) Add a layer of census block groups. Another requirement for the new park is that it be located in a low-income neighborhood. To symbolize an income attribute, we have to add another layer to the map. (It's not that income data isn't reported at the census tract level, it's just that our particular *tracts* layer doesn't happen to include it.)

❶ Open the Catalog window.

The same census folder that contains *tracts* also has a dataset called *block_groups*.

❷ Drag and drop *block_groups* on the map.

In the Table of Contents, the new layer is added above the other polygon layers. Like the *tracts* layer, the *block_groups* layer covers Los Angeles County. And like the census tracts, the block group polygons resemble a complicated jigsaw puzzle. Block groups are another

Census Bureau statistical unit: they're smaller than tracts and nest inside them by design. We'll talk more about census geography in Lesson 2.

C Open the attribute table for the *block_groups* layer and scroll across the attributes.

Most of these attribute names are *really* cryptic. One of the last fields in the table is called MEDHINC_CY (Figure 1-81). For now, take it on faith that this is median household income.

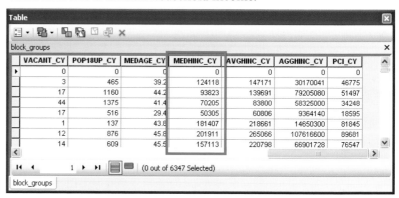

Table

block_groups

	VACANT_CY	POP18UP_CY	MEDAGE_CY	MEDHINC_CY	AVGHINC_CY	AGGHINC_CY	PCI_CY
▶	0	0	0	0	0	0	0
	3	465	39.2	124118	147171	30170041	46775
	17	1160	44.2	93823	139691	79205080	51497
	44	1375	41.4	70205	83800	58325000	34248
	17	516	29.4	50305	60806	9364140	18595
	1	137	43.8	181407	218661	14650300	81845
	12	876	45.8	201911	265066	107616600	89681
	14	609	45.5	157113	220798	66901728	76547

I◄ ◄   1   ► ►I   (0 out of 6347 Selected)

block_groups

Figure 1-81

D Right-click the MEDHINC_CY field name and choose Statistics.

The Statistics of block_groups window tells you that the lowest value in the field is 0 and the hightest is 347,238. Median value is a midpoint. For each block group, half the households earn more than the median income and half earn less.

E Close the Statistics of block_groups window.

F Close the attribute table.

15) Symbolize census block groups by median household income. If we symbolize the *block_groups* layer with graduated colors, we won't be able to evaluate income and population density at the same time. Instead, we'll represent each block group's median household income as a point drawn inside the block group polygon. The point sizes will be graduated according to the income value.

A Open the layer properties for the *block_groups* layer and click the Symbology tab.

B In the Show box, click Quantities. Under Quantities, click Graduated Symbols (Figure 1-82).

C In the Fields area, click the Value drop-down list and choose MEDHINC_CY.

D Click the Classes drop-down arrow and choose 3.

E Click Classify to open the Classification dialog box.

F In the Break Values box to the right of the histogram, click the the first class break point (56281) to make it editable.

Figure 1-82

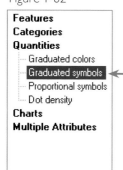

Features
Categories
Quantities
    Graduated colors
    Graduated symbols
    Proportional symbols
    Dot density
Charts
Multiple Attributes

ⓖ Replace the highlighted value with **50,000** and press Enter.

ⓗ Replace the highlighted value for the second break point (115630) with **100,000**.

ⓘ Click in some white space in the Histogram window to stop editing class breaks (Figure 1-83). Click OK on the Classification dialog box.

ⓙ In the Layer Properties dialog box, click Template (underneath the Classify button) to open the Symbol Selector.

ⓚ In the scrolling box of symbols, click Circle 2 to select it.

ⓛ Click the Color button and change its color to Tourmaline Green (Figure 1-84).

Figure 1-83

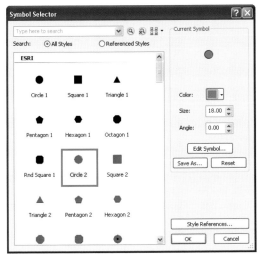

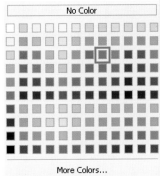

Figure 1-84

ⓜ Click OK on the Symbol Selector.

This sets the color and shape of the symbols that will be used to represent income values.

ⓝ Click Background (underneath Template) to open the Symbol Selector again.

ⓞ Change the fill color to No Color, then change the outline color to No Color. Click OK.

This makes the block group polygons themselves invisible. All we'll see are the income dots spread around the map.

ⓟ In the Symbol Size boxes, replace the "from" value with **8** and the "to" value with **24**.

ⓠ In the Label column, click on the first label (0 - 50000) to make it editable. Type **Low** and press Enter.

ⓡ Replace the second label with **Medium**. Press Enter.

S Change the third label to **High**. Click outside the edit box to commit the edit.

T Compare your settings to Figure 1-85. Click Apply, then move the Layer Properties dialog box out of the way.

Figure 1-85

Figure 1-86

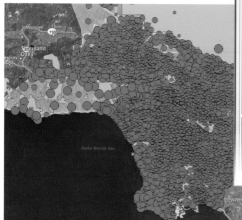

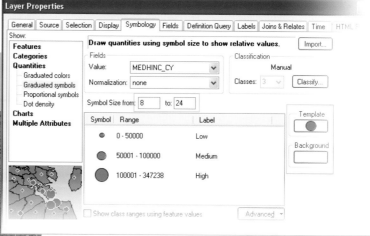

Your screen may look different depending on the size of your computer screen and ArcMap window, but the effect should be similar.

At the present scale, the symbols overwhelm the map (Figure 1-86).

16) Set a scale range for the *block_groups* layer. Earlier, we set a scale range for the *Parkland* labels so that they wouldn't display when we zoomed out past a certain scale. Here, we'll do the same thing for the *block_groups* layer.

A On the Layer Properties dialog box, click the General tab.

B In the Scale Range area, click the "Don't show layer when zoomed" option.

C Click in the "Out beyond" box and type **100,000** (Figure 1-87), then click Apply.

Figure 1-87

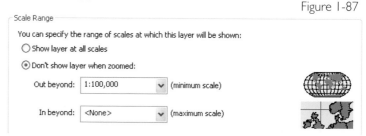

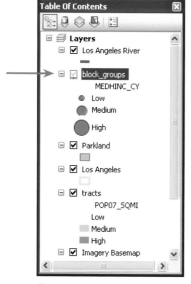

Figure 1-88

D Move the Layer Properties dialog box out of the way.

The symbols disappear from the map. In the Table of Contents, the layer's check box is gray (Figure 1-88). This means that the layer is visible, but not at the present map scale.

ⓔ In the Layer Properties dialog box, click the Display tab.

ⓕ In the Transparent % box, type **20**. Click OK.

ⓖ Zoom to the Dodger Stadium bookmark.

Now we can start to get a general sense of household income and population density along the river, and look at these variables in relation to park locations (Figure 1-89).

Figure 1-89

17) **Search for likely park areas.** Clearly, we're taking an incomplete initial look at a more complex problem. We haven't considered all the requirements (for example, the presence of children). We're not making any exact measurements of distance. Our data classifications are casual: we don't yet have a good reason to say what values should count as high population density or low median household income in the context of our project. Nevertheless, we can form some meaningful impressions. We won't be able to say of an area that a park should *definitely* go there, but we might be able to identify likely and unlikely areas. Later, it will be interesting to see how well these impressions are borne out by rigorous analysis.

ⓐ On the Standard toolbar, in the Map Scale box, highlight the current value. Type **50,000** and press Enter.

ⓑ Pan south to where the river crosses the city boundary.

We'll follow the river to its source, marking good areas along the way. A really good area would have these properties:

- High population density (dark red)
- Low median household income (small green dot)

- No existing park nearby
- Close to the river

We can rule out areas with parks, even if they're otherwise good. We can also rule out all areas that aren't within a mile or so of the river.

ⓒ Pan slowly north.

Take your time—there's a lot to look at. Soon, just north of the 101 Freeway, and about a mile and a half southeast of Dodger Stadium, you'll come to a dark red tract with a small green dot and no park nearby. This area looks ideal, so let's mark it.

ⓓ From the main menu, choose Customize → Toolbars → Draw.

The Draw toolbar is added (Figure 1-90).

Figure 1-90

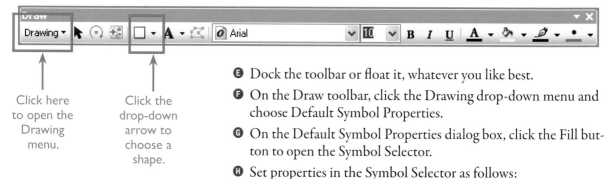

Click here to open the Drawing menu.

Click the drop-down arrow to choose a shape.

ⓔ Dock the toolbar or float it, whatever you like best.

ⓕ On the Draw toolbar, click the Drawing drop-down menu and choose Default Symbol Properties.

ⓖ On the Default Symbol Properties dialog box, click the Fill button to open the Symbol Selector.

ⓗ Set properties in the Symbol Selector as follows:
- Set the fill color to No Color.
- Set the outline width to 3.
- Set the outline color to Black.

ⓘ Click OK on the Symbol Selector.

ⓙ Compare your Default Symbol Properties dialog box to Figure 1-91, then click OK.

ⓚ On the Draw toolbar, click the drop-down arrow next to the Rectangle tool. From the list of tools, choose the Ellipse tool.

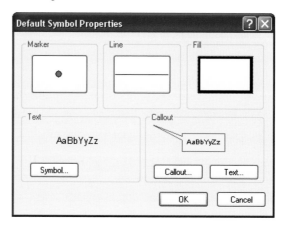

Figure 1-91

◐ On the map, click and drag a box over the tract of interest. Release the mouse button to draw the ellipse (Figure 1-92).

You can reshape the ellipse by dragging its selection handles. You can move it by clicking and dragging it. The exact size and position of the shape doesn't matter as long as you mark the general area.

◍ Click off the graphic to unselect it.

Unfortunately, this promising area won't meet our final park requirements. The Los Angeles County Men's Central Jail is located here, and although the population density is high, there aren't many children. Local knowledge and familiarity with your study area are an important factor in anticipating and evaluating analysis results.

Figure 1-92

Graphics

Map graphics are different from features. Conceptually, features represent geographic objects while graphics tend to be visual aids. Technically, graphics are usually stored in map documents as annotation, not in feature classes on disk. (You'll learn more about this in Lesson 8.) They aren't maintained as layers in ArcMap, nor do they have attribute tables, but they do have group properties that can be managed within a data frame. Features can be converted to graphics and graphics can be converted to features.

18) Continue searching for park areas. We'll keep following the river and looking for likely places for a park.

④ Pan north.

Northeast of Dodger Stadium and east of the river, is a high-density, low-income area. It's not right next to the river, and there is a park in the general area, but let's mark this spot anyway.

⑱ On the Drawing menu, click the Ellipse tool and draw an ellipse around the area (Figure 1-93).

The shape of the ellipse doesn't fit the area very well.

⑥ On the Draw toolbar, click the Rotate button ⟳ .

⑩ Place the mouse pointer over the graphic, then click and drag the mouse slightly to rotate the graphic. Orient the ellipse to fit the area better.

⑤ On the Draw toolbar, click the Select Elements tool ➤ .

⑥ Resize and reshape the ellipse by dragging selection handles.

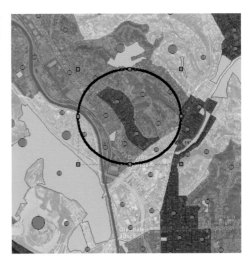

Figure 1-93

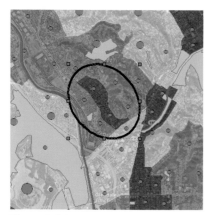

Figure 1-94

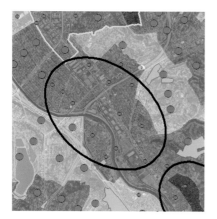

Figure 1-95

You may need to switch back and forth between the Rotate and Select Elements tool to make the ellipse look the way you want (Figure 1-94). It's not art, so don't spend too much time on it.

G Unselect the graphic and pan a little farther.

The area around the intersection of the 5 and 2 freeways, northeast of the Silver Lake Reservoir, might also be good. The population density is medium, but the income is low and there aren't many parks.

H Add a graphic in this area of a shape that seems right to you.

Figure 1-95 is an example, but your own estimate may be just as good or better.

I Unselect the graphic and keep panning along the river.

As you navigate around Griffith Park, remember that the yellow line marks the city limits. The areas north and east of the park aren't part of Los Angeles and shouldn't be considered.

J Continue to the end of the river, drawing ellipses wherever you think they should go.

Once you get past Griffith Park, you may find that likely areas are scarce. But remember, there's no right or wrong answer for this exercise.

K After adding the last ellipse, from the main menu, choose Edit → Select All Elements.

L Again from the main menu, choose Edit → Zoom to Selected Elements.

M Click anywhere on the map to unselect the graphic elements (Figure 1-96).

N Save the map.

Your guesses will probably be different, which is fine.

Figure 1-96

19) Export the map. We'll export this view of the map to an image in TIFF format. When we finish the analysis in Lesson 6, we'll add the image and compare our guesses to the actual results.

A From the main menu, choose File → Export Map.

B In the Export Map dialog box, in the Save in box at the top, navigate, if necessary, to the MapsAndMore folder.

C In the Save as type box, click the drop-down arrow and choose TIFF (*.tif).

D Replace the file name with **Lesson1Predictions**.

E Under Options, click the Format tab. At the bottom, check the box to write GeoTIFF tags.

GeoTIFF tags embed coordinate system information into the image file. The output will be an image that can be used in documents like presentations or reports, but that will also align spatially with layers in ArcMap.

F Compare your Export Map dialog box to Figure 1-97, then click Save.

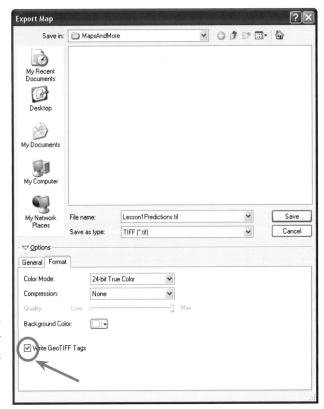

Figure 1-97

The Export Map dialog box closes by itself.

20) Exit ArcMap. Toolbars that you leave on the interface will remain there in subsequent sessions. There's nothing wrong with that, but too many toolbars can become a distraction.

A From the main menu, choose Customize → Toolbars → Draw.

The Draw toolbar is removed.

B Remove the Effects toolbar in the same way.

C From the main menu, choose File → Exit.

At this point, you already have a basic working knowledge of ArcMap. You know the problem and you have a sense of the study area. You've worked with some of the available data for the project, but you haven't taken a careful, systematic look at it. That will be the focus of the next lesson.

Symbology methods

The default **Single Symbol** method (under Features) assigns the same symbol to all features in a layer. No attribute is symbolized.

All U.S. states have the same fill and outline color.

Categorical methods are used with descriptive (usually text-based) attributes. Features that have the same descriptive value get the same symbol.

Roads are symbolized by a road type attribute. Different line widths are used for freeways, highways, and streets.

Chart methods compare two or more numeric attributes.

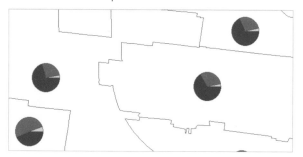

Occupancy attributes are symbolized with pie charts. The percentage of owner-occupied homes is shown in red, renter-occupied homes in blue, and unoccupied homes in yellow. Each chart represents a different city.

Quantitative methods are used with numeric attributes. Features with similar numeric values are grouped and given the same symbol.

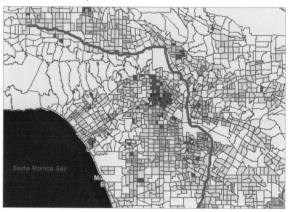

Census tracts are symbolized by a population density attribute. Low-density tracts are yellow, medium-density are orange, and high-density are brown.

The **Quantity by category** method (under Multiple Attributes) symbolizes a descriptive and a numeric attribute at the same time.

The descriptive attribute is parcel vacancy. Vacant parcels are orange, developed parcels are purple. The numeric attribute is assessed land value. Small orange dots in vacant parcels mean the assessed value is relatively low; large dots mean it's high.

Lesson

2 Preview data

WE NEED THE RIGHT DATA TO SOLVE

the problem. We explored the study area in Lesson 1; now we can proceed more systematically. What data do we have? How useful is it? Is there data that we need but don't have? Has the problem been stated clearly enough for us to know what data we need?

Acquiring, evaluating, and organizing data is a big part of an analysis project. This book doesn't fully re-create the complexity of the real world because we've provided all the basic data required. On the other hand, much of the data isn't project-ready, and that does reflect the real world of GIS.

The first thing we'll do in this lesson is draw up a planning document to help us keep our tasks in focus. We'll use this document to list the guidelines for the new park and translate them into specific needs for spatial and attribute data.

After we itemize our data requirements in general terms (park data, river data, and so on), we'll take stock of our source data and investigate its spatial and attribute properties. We'll also familiarize ourselves with metadata, which is the data we have about our data. Before you decide to use a particular dataset, you may want to know things like who made the data, when, and to what standard of accuracy.

Once we have a better working knowledge of our data, we'll reframe the problem statement. GIS is a quantitative technology: we can't analyze a problem until it's been stated in measurable terms. As a GIS analyst, the first thought that should cross your mind when you're asked to find a location for a park site near the river is, "What do you mean by 'near?'" Wherever we find the city council's guidelines to be vague, we'll replace them with hard numbers.

You are here:

What you'll do in this lesson:

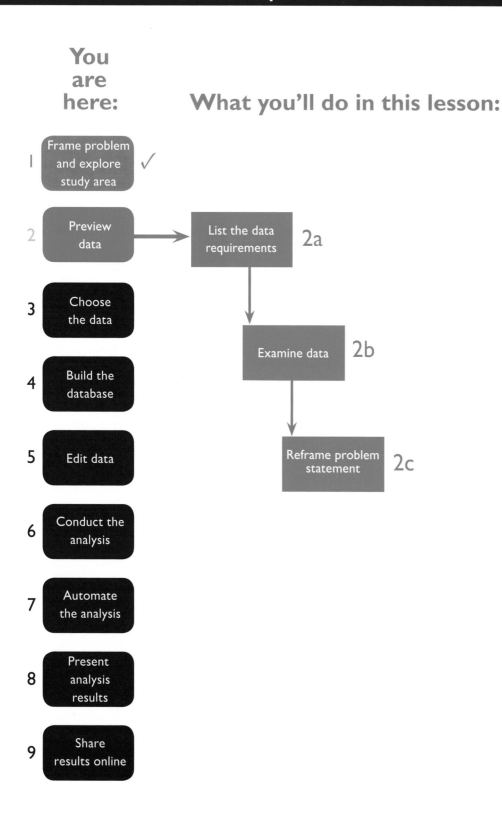

1. Frame problem and explore study area ✓

2. Preview data → List the data requirements — 2a

3. Choose the data

   Examine data — 2b

4. Build the database

5. Edit data

   Reframe problem statement — 2c

6. Conduct the analysis

7. Automate the analysis

8. Present analysis results

9. Share results online

Exercise 2a: List the data requirements

We need to relate the guidelines for the new park to data requirements for the project.

1) **Open the data requirements table.** We've made a table in advance to keep track of our requirements. It's an informal document, but it will still be helpful.

- Ⓐ Open Windows Explorer and navigate to C:\UGIS\ParkSite\MapsAndMore.
- Ⓑ Double-click the file *DataRequirementsTable.doc* to open it in Microsoft Word (Figure 2-1).
  - ▷ If you don't have Microsoft Word, open the .rtf version of the document in another application, or print the .pdf version and fill it out with a pencil.

Figure 2-1

#	REQUIREMENT	DEFINED AS	SPATIAL DATA	ATTRIBUTE DATA	DATA SET	PREPARATION
1	land parcel		parcels			
2	vacant					
3						
4						
5						
6						
7						
8						
9						
10						
11						
12						
13						
14						
15						

2) **List the requirements.** In this step, we'll review the city council's guidelines (see next page) and describe in a general way the data needed to satisfy them. We'll get to the specifics of choosing datasets in Lesson 3.

Pieces of land, legally surveyed, described, and recorded with a county or other administrative body are called parcels.

The first guideline was to find a vacant piece of land at least one acre in size. We can break this down into three requirements:

- land parcel
- vacancy
- size

The requirement for a land parcel is already listed in the table. We need spatial data representing parcels so that we can see candidate sites on the map.

The second requirement is vacancy, which is a characteristic, or attribute, of a parcel. Vacancy is usually included in the more general heading of land use, which describes the purpose the parcel is being put to (commercial, residential, industrial, and so on).

Park siting guidelines
• On a vacant land parcel one acre or larger • Within the LA city limits • As close as possible to the LA River • Not in the vicinity of an existing park • In a densely populated neighbor-hood with lots of children • In a low-income neighborhood • Serving as many people as possible

Figure 2-2

Ⓐ In Row 2 of the table, under ATTRIBUTE DATA, type (or write) **`land use`**.

The third requirement is that the park be one acre or larger. Like vacancy, acreage is an attribute, although it is one that can be calculated by the software. Since ArcMap can convert one unit of area to another, we don't have to start with acres—any measurement of parcel size will suffice.

Ⓑ In Row 3, under REQUIREMENT, type **`one or more acres`**. Under ATTRIBUTE DATA, enter **`area`**.

The second guideline is that the park be within the LA city limits (Figure 2-2). This sounds like spatial data and we'll treat it that way, at least for now. (It could be an attribute, too, because a field in a table might store the name of the city where each parcel is recorded.)

Ⓒ Fill in Row 4 as you think it should look, then check Figure 2-3.

#	REQUIREMENT	DEFINED AS	SPATIAL DATA	ATTRIBUTE DATA
1	land parcel		parcels	
2	vacant			land use
3	one or more acres			area
4	within city limits		cities	
5				
6				

Figure 2-3

The third guideline is that the park be as close as possible to the Los Angeles River.

Ⓓ In Row 5, for the requirement, put **`near LA river`**. Under SPATIAL DATA, put **`rivers`**.

Distance to the river can be thought of as an attribute of parcels, but it's not clear that we need it as data. We might just make distance measurements on the map once we have some candidate parcels that meet our other requirements. If we decide later to store distance as a parcel attribute, it's something ArcMap can calculate for us.

The fourth guideline is to put the park away from existing parks.

Ⓔ Fill out Row 6 as you think it should look.

The fifth guideline also needs to be broken down. We need

- A neighborhood (spatial data) with
- High population density (attribute data) and
- Lots of children (attribute data)

Neighborhoods tend not to have formal boundaries, so we're probably not going to find them as such in a feature class. As a proxy, or substitute, we'll use a set of small, standardized areas defined by the U.S. Census Bureau: either the tracts or block groups we looked at in Lesson 1.

ⓕ In Row 7, enter **in a neighborhood** as the requirement. Enter **census unit** for the spatial data.

ⓖ In Row 8, enter **densely populated** for the requirement and **population density** for the attribute data.

ⓗ In Row 9, enter **lots of kids** for the requirement. For the attribute data, enter **age**.

The sixth guideline is that the park go in a low-income neighborhood. We don't need to repeat the spatial requirement for a neighborhood.

ⓘ In Row 10, enter **low income** for the requirement, and **income** for the attribute data.

The last guideline is to serve as many people as possible. For this we need a population attribute.

ⓙ In Row 11, enter **serving the most people** as the requirement and **population** as the attribute data (Figure 2-4).

#	REQUIREMENT	DEFINED AS	SPATIAL DATA	ATTRIBUTE DATA	DATA SET
1	land parcel		parcels		
2	vacant			land use	
3	one or more acres			area	
4	within city limits		cities		
5	near LA river		rivers		
6	away from parks		parks		
7	in a neighborhood		census unit		
8	densely populated			population density	
9	lots of kids			age	
10	low income			income	
11	serving the most people			population	
12					

Figure 2-4

Eventually, we want to make a map of potential sites, and we may need some data just for cartographic purposes. For example, political boundaries and roads put a map in a familiar context. Physical relief gives it texture and imagery can provide realism.

ⓚ In Rows 12 to 15, enter **f inal map** for the requirement. Under SPATIAL DATA, list the examples just mentioned (Figure 2-5).

ⓛ Save and minimize the table. We'll continue to use it.

#	REQUIREMENT	DEFINED AS	SPATIAL DATA	ATTRIBUTE DATA	DATA SET	PREP
1	land parcel		parcels			
2	vacant			land use		
3	one or more acres			area		
4	within city limits		cities			
5	near LA river		rivers			
6	away from parks		parks			
7	in a neighborhood		census unit			
8	densely populated			population density		
9	lots of kids			age		
10	low income			income		
11	serving the most people			population		
12	final map		political boundaries			
13	final map		roads			
14	final map		relief			
15	final map		imagery			

Figure 2-5

Exercise 2b: Examine data

Now let's see what data we actually have on hand. To do this, we'll work in ArcCatalog, the data management application of ArcGIS Desktop. In Lesson 1, you used the Catalog window, which is basically ArcCatalog tailored to work inside ArcMap. The Catalog window is great for going back and forth between your map and your data (which is probably most of the time). In this lesson, however, we won't be doing any mapping, and the stand-alone ArcCatalog application will be a little more convenient to work with.

1) **Start ArcCatalog.** We'll open the ArcCatalog application.

Ⓐ Click the Windows Start button, then choose All Programs → ArcGIS → ArcCatalog 10 (Figure 2-6).

▷ If you have a shortcut on your desktop, you can double-click the ArcCatalog 10 icon 🗐 instead.

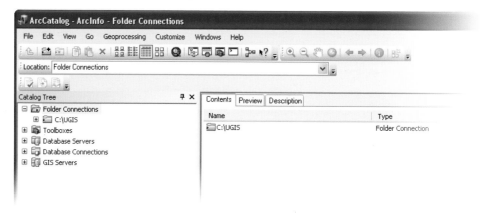

Figure 2-6

The Catalog Tree, on the left, is the counterpart of the Catalog window in ArcMap. On the right are three tabs.

- The Contents tab shows file information.
- The Preview tab lets you explore spatial and attribute data without opening a map document.
- The Description tab displays metadata (data about data).

2) **Show file extensions.** By default, ArcGIS hides file extensions. You may find it convenient to see them.

Ⓐ From the main menu, choose Customize → ArcCatalog Options.

Ⓑ If necessary, click the General tab.

Ⓒ Near the bottom of the tab, uncheck the Hide file extensions box. Click OK in the ArcCatalog Options dialog box.

Representing the real world as data

How would you create a system to organize and manage the bewildering variety of stuff in the world? One way is to think of geography in terms of discrete objects. If the world is a huge set of objects, you can sort them on the basis of similarities. Shape is one sorting principle: every object can be represented as either a point, a line, or a polygon. Theme is another principle: every object can be classified as a city, a river, a park, or something else. Using these principles, we end up with heaps of different things: parks represented as polygons, rivers represented as lines, and so on. Within each heap, objects have their own shapes (all parks are polygons, but each park's polygon is different) and/or their own locations. Each object's place in the world is staked out by a set of spatial coordinates. Objects are also differentiated by a unique collection of facts: the individual names, measurements, and descriptions that express what we know about them. These are the basic ideas behind the concept of a feature class: a set of thematically related objects with the same type of shape, with real-world locations, and with a common set of descriptive characteristics (attributes) stored in a table. Each object is called a feature. It's a piece of geometry that draws on a map with a certain shape and location. It's also a record (row) in a feature attribute table that stores descriptive information about it. This conceptual model of geography as discrete objects is called the vector data model.

These icons		Mean
⊡	⊡	Point feature class
⊣	⊣	Line feature class
▧	▧	Polygon feature class

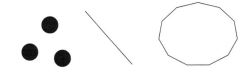

The difference between the blue and green icons is the file format. The blue ones are geodatabase feature classes, and the green ones are shapefile feature classes. Geodatabase feature classes are stored in geodatabases 🗄 and sometimes in feature datasets 🗗 within a geodatabase. Shapefiles are stored in ordinary file folders 📁. Both types of feature class model geography in basically the same way (point, line, polygon), but the geodatabase format is newer and has advanced capabilities that shapefiles lack.

The vector model has conceptual limitations. It's not a very intuitive way to to represent aspects of geography like elevation or temperature that don't really have shapes or boundaries and that cover the world everywhere. A second way to think of geography is in terms of surfaces. A surface is partitioned into cells, or pixels, in which each cell represents a unit of area, like a square meter, and stores a numeric measurement or estimate. This model of geography is called the raster data model. It is used for imagery and all kinds of continuously distributed phenomena.

These icons		Mean
▦	▦	Raster dataset

These icons		Mean
▦	▤	Stand-alone table

The blue icon means the raster is stored in geodatabase format. The green icon means it's stored in some other file format—commonly .jpg or .tif for images. Raster datasets usually don't have tables or else have very simple ones.

A GIS can also store tables of information that stand alone and don't belong to a feature class. The blue icon means the table is stored in geodatabase format. The green icon means it's stored in another format, such as dBASE® format.

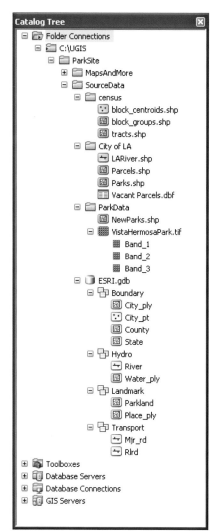

Figure 2-7

3) **Expand the folder connection.** In the Catalog Tree, you should see a folder connection to C:\UGIS. You made this connection from the Add Data button in ArcMap in Lesson 1, and ArcCatalog has inherited it.

Ⓐ In the Catalog Tree, expand the C:\UGIS folder connection by clicking its plus sign.

▷ If you don't have this folder connection, make it now. On the Standard toolbar, click the Connect To Folder button 📇. Browse to C:\UGIS and click OK.

Ⓑ Expand the ParkSite folder and the SourceData folder below it.

Ⓒ Under SourceData, expand everything you can.

It's a long list of items (Figure 2-7). With the Catalog Tree docked in ArcCatalog, you may have to scroll down or maximize the application to see all the data. Each item is a piece of geographic data or a data container. The icons signify the type of data, as illustrated in the topic *Representing the real world as data.*

4) **Survey the data.** Under the SourceData folder are three folders 📁 and a geodatabase 🗄.

- The census folder contains three feature classes of census data in shapefile format.
- The City of LA folder contains three shapefiles and a stand-alone table in dBASE format.
- The ParkData folder contains a shapefile and a raster dataset.
- The ESRI geodatabase contains ten feature classes in geodatabase format. The feature classes are thematically organized in containers called feature datasets 🗂.

In the next several steps, we'll preview a lot of this data to make sure we have the features and attributes we listed in the data requirements table.

Finding data

In a real GIS project, finding good data is a big part of your job. An ArcGIS Desktop license gives you access to the Esri Data & Maps media kit, which includes over 160 presymbolized vector data layers for North America, Europe, and the world. From ArcGIS Desktop, you can also connect to map services (remotely-hosted data) such as the cartographic basemaps you'll use in this book. These basemaps, along with many other map services and downloadable datasets, are part of a large and growing repository of mapping resources called ArcGIS Online. You can search ArcGIS Online for map content either directly from ArcMap or from the ArcGIS.com website. Spatial data is also widely available from government agencies, educational institutions, and commercial vendors. All of these sources may supplement data collected and managed by your own organization.

5) Preview parcels. First, we'll preview the parcels data, which comes from the City of Los Angeles.

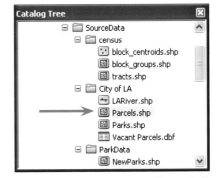

Figure 2-8

A In the Catalog Tree, under the City of LA folder, click *Parcels.shp* to highlight it (Figure 2-8).

The icon ▦ signifies a polygon feature class in shapefile format. On the Contents tab, you see a thumbnail graphic of the data.

B Click the Preview tab.

This gives you a larger and interactive view of the data. The previously disabled tools on the Geography toolbar (Figure 2-9) can now be used.

Figure 2-9

The Geography toolbar

C On the Geography toolbar, click the Zoom In tool ⊕.

D Click and drag a box, like the one shown in Figure 2-10, anywhere on the data.

E Zoom in farther until you can easily distinguish individual features, as in Figure 2-11.

This is the spatial data we need, showing individual parcel boundaries. By default, previewed features display with simple and uniform symbology. Polygon features are yellow with gray outlines.

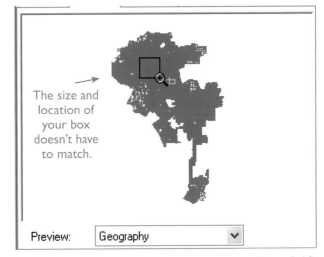

The size and location of your box doesn't have to match.

Figure 2-10

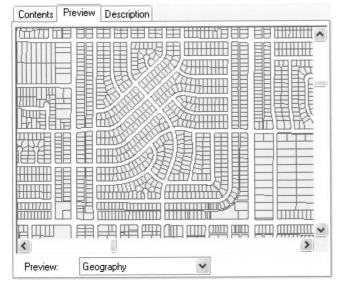

Figure 2-11

G On the Geography toolbar, click the Identify tool **ⓘ**. Click on any parcel to identify it. (Figure 2-12).

   ▷ The Parcels dataset is big, so the response may be a little slow.

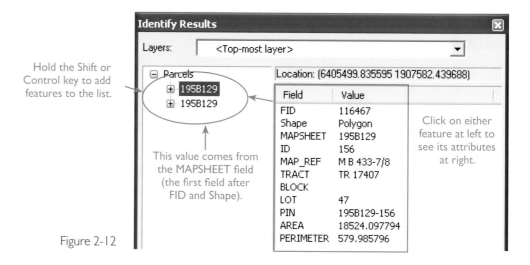

Hold the Shift or Control key to add features to the list.

This value comes from the MAPSHEET field (the first field after FID and Shape).

Click on either feature at left to see its attributes at right.

Figure 2-12

Depending on exactly where you click, you may identify one or more features. On the left side of the Identify Results window, the identified parcel is listed. On the right side are the parcel's attributes: mostly codes and identification numbers.

The vacancy attribute we need isn't here, but we do have an AREA field in unspecified units. Once we find out what the units are (Lesson 3), we can convert them to acres. (An acre is about the same size as an American football field, but if that doesn't mean anything to you, Figure 2-13 and 2-14 provide some context.)

Ten acres

Figure 2-13

Figure 2-14

One acre

G Close the Identify Results window.

6) Preview the table of vacant parcels. Since we don't have a vacancy attribute in the parcels table, we'll look for it elsewhere.

A In the Catalog Tree, under the City of LA folder, click *Vacant Parcels.dbf*.

This is a stand-alone table. It has information about parcels, but no association with spatial features. Each of the 44,462 records represents a vacant parcel (Figure 2-15).

B Scroll across the attributes.

OID is a software-managed record identifier. PIN (Parcel Identification Number) and APN (Assessor Parcel Number) are user-managed identifiers. Several other attributes follow. In the LANDUSE field, all values end in "V," the code for vacancy.

C Click the Description tab (Figure 2-16).

The Description tab shows metadata, or data documentation. Complete metadata, which we'll access later in the lesson, includes a lot of elements. What you see here is called the Item Description, an overview of the dataset.

D Scroll down through the Item Description.

From the top, you see:

- The dataset name (Vacant Parcels) and file type (dBASE Table)
- A thumbnail graphic (not available for stand-alone tables)
- Tags that make the data searchable
- A summary of the intended use of the data
- A description of the data content
- Credits
- Access and use limitations

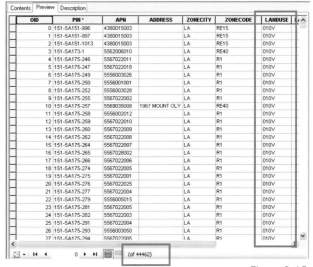

Figure 2-15

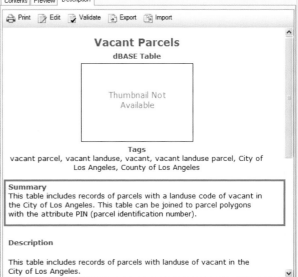

Figure 2-16

In this case, the summary contains a key piece of information. The records in this table can be joined, or connected, to features in the *Parcels* feature class using the PIN attribute, which is common to both tables (look back at Figure 2-12). Records with matching PIN numbers in the two tables can share attributes. This means that parcel features can be queried, selected, and symbolized on the LANDUSE attribute in the *Vacant Parcels* table. In other words, we have a way to determine which parcels on the map are vacant. We'll get to this operation, called a table join, in Lesson 4.

7) **Preview cities.** Row 4 of the requirements table lists cities as needed spatial data. We have this data—we used it in Lesson 1, when we made a definition query on Los Angeles.

Ⓐ Click the Preview tab.

Ⓑ In the Catalog Tree, under ESRI.gdb, under Boundary, click the *City_ply* geodatabase feature class to highlight it (Figure 2-17).

The preview changes from Table to Geography. The window is all yellow because you're still zoomed in to where you were previewing the *Parcels* feature class.

Ⓒ On the Geography toolbar, click the Full Extent button .

You zoom out to a view of all the features in the dataset.

Ⓓ In the Catalog Tree, still under the Boundary feature dataset, click *City_pt*.

The main difference between the two feature classes is that *City_pt* represents cities as points rather than polygons. (Hence the names *City_pt* and *City_ply*.) There's also a difference in spatial extent because *City_pt* includes a feature for a place called Attu Station in Alaska, which is so far west that it appears in the east (Figure 2-18).

Figure 2-17

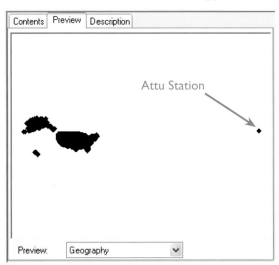

Figure 2-18

E Zoom in on the data to see individual features.

Why would the same features, such as cities, be represented with different types of shapes? It's because they're appropriate for maps of different scale. For a national map, you would show cities as points. For a local map, or maybe a *very* local map, you would show them as polygons. For our requirement, which is to make sure that parcels lie inside the city limits, we need features that have boundaries. The polygon feature class is obviously the right one.

F Click the Full Extent button.

8)   Preview the LA River. In Row 5 of the requirements table, we have rivers as needed spatial data. In Lesson 1 we added the *River* feature class from the ESRI geodatabase. We also have another dataset to look at: the *LARiver* shapefile in the City of LA folder.

A Preview the geography of the *LARiver* shapefile.

Line features preview in dark blue. It looks like there's just one river in this feature class.

B At the bottom of the Preview tab, click the drop-down arrow and change the preview from Geography to Table.

C Scroll down to the last record in the table.

In fact, the Los Angeles River *is* the only river in the feature class, but it's composed of 265 separate features (the FID, or feature ID, of the first record is 0). Why so many? As before, the answer has to do with attributes.

D Scroll across the table to see the attributes, then scroll up and notice the different values in the CAPACITY, DISCHARGE, and PROTECTION fields.

For our purposes, it doesn't really matter what these attributes mean. The point is that if the river were represented as a single feature, it would also have just one row in the attribute table. That would mean that only a single value could be stored for each attribute—fine for the river name (which doesn't change), but a problem for anything we might want to measure or describe at different locations along the river: flow, depth, water chemistry, navigability, or anything else. The creators of this data wanted to gather facts about the river at different places. To do that, they had to define the river as a spatially connected series of individual features.

We don't share that need. In this case, we want the spatial data, not the attributes. If we end up using this dataset in our analysis (rather than the *River* feature class in ESRI.gdb), we'll probably reduce the 265 features to one.

9) **Preview parks.** In Row 6 of the requirements table, we need spatial data representing parks. We already know we have parks data: in Lesson 1, we symbolized and labeled the *Parkland* feature class. There's also a shapefile called *Parks* in the City of LA folder.

- **A** Set the preview back to Geography.
- **B** In the Catalog Tree, click *Parks.shp*.
- **C** Preview the geography, then the table (Figure 2-19).

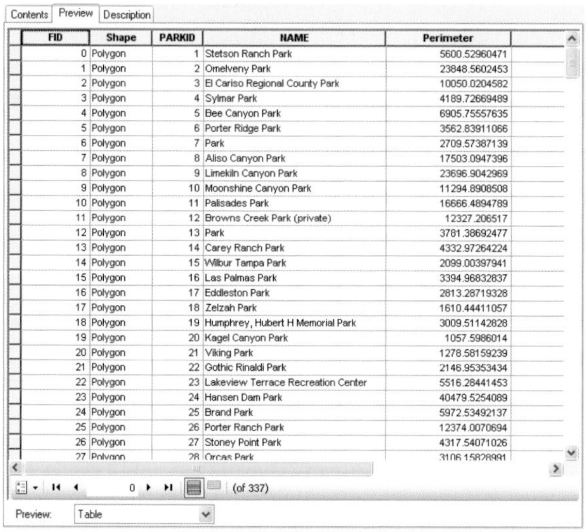

FID	Shape	PARKID	NAME	Perimeter
0	Polygon	1	Stetson Ranch Park	5600.52960471
1	Polygon	2	Omelveny Park	23848.5602453
2	Polygon	3	El Cariso Regional County Park	10050.0204582
3	Polygon	4	Sylmar Park	4189.72669489
4	Polygon	5	Bee Canyon Park	6905.75557635
5	Polygon	6	Porter Ridge Park	3562.83911066
6	Polygon	7	Park	2709.57387139
7	Polygon	8	Aliso Canyon Park	17503.0947396
8	Polygon	9	Limekiln Canyon Park	23696.9042969
9	Polygon	10	Moonshine Canyon Park	11294.8908508
10	Polygon	11	Palisades Park	16666.4894789
11	Polygon	12	Browns Creek Park (private)	12327.206517
12	Polygon	13	Park	3781.38692477
13	Polygon	14	Carey Ranch Park	4332.97264224
14	Polygon	15	Wilbur Tampa Park	2099.00397941
15	Polygon	16	Las Palmas Park	3394.96832837
16	Polygon	17	Eddleston Park	2813.28719328
17	Polygon	18	Zelzah Park	1610.44411057
18	Polygon	19	Humphrey, Hubert H Memorial Park	3009.51142828
19	Polygon	20	Kagel Canyon Park	1057.5986014
20	Polygon	21	Viking Park	1278.58159239
21	Polygon	22	Gothic Rinaldi Park	2146.95353434
22	Polygon	23	Lakeview Terrace Recreation Center	5516.28441453
23	Polygon	24	Hansen Dam Park	40479.5254089
24	Polygon	25	Brand Park	5972.53492137
25	Polygon	26	Porter Ranch Park	12374.0070694
26	Polygon	27	Stoney Point Park	4317.54071026
27	Polygon	28	Orcas Park	3106.15828991

0 (of 337)

Preview: Table

Figure 2-19

It looks from the geography preview like the extent of the features corresponds to the city of Los Angeles. We can't be sure until we add the data to the map. The table preview shows us that there are 337 records (features) and just a few attributes.

Software-managed attributes

Every shapefile feature class has FID and Shape attributes that are created and managed by the software. The FID (feature identification) attribute stores a unique number for each feature. The Shape attribute stores the geometry type. Behind the scenes, it links each feature to coordinates that define its spatial location. Measurement attributes, such as length and area, can be calculated for shapefiles, but if the values change—because of a spatial edit, for example—the software doesn't update them automatically.

A geodatabase feature class has up to four software-managed attributes. Like a shapefile, it has a feature identifer (called OBJECTID instead of FID) and a Shape attribute. The Shape_Length attribute stores the length of line and polygon features. This attribute doesn't exist for point features. The Shape_Area attribute stores the internal area of polygon features. It doesn't exist for point or line features. Shape_Length and Shape_Area are automatically kept up-to-date by the software.

D Set the preview back to Geography.

E In the Catalog Tree, under the ParkData folder, click *NewParks.shp*.

F Preview its geography and table.

This shapefile has just two features. One is Los Angeles (State Historic Park) and the other is Rio de Los Angeles (State Recreation Area). Note the absence of length and area attributes. We can make them if we want—for example, the *Parks* shapefile has them—but they don't exist by default.

G Click the Description tab (Figure 2-20) and read the summary.

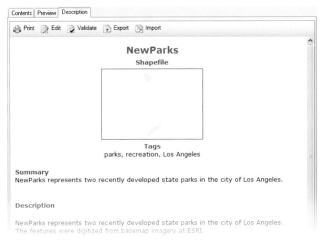

Figure 2-20

The data represents two newly developed parks. In Lesson 3, when we choose a parks feature class for the analysis, we'll have to make sure that it includes these two parks.

The ParkData folder also contains a raster dataset.

H Click the Preview tab.

I In the Catalog Tree, click *VistaHermosaPark.tif*.

The geography preview shows an aerial photograph of a park under development (Figure 2-21). Located just north of downtown, Vista Hermosa Park has since been completed. It's not included in the *NewParks* shapefile, however, so it's one more thing to remember when we choose the park data in Lesson 3.

There's no table to preview. The information associated with an image pixel is just its color value.

J On the Geography toolbar, click the Identify tool ⓘ .

K Identify a location anywhere on the image to see the RGB value of the identified pixel (Figure 2-22).

L Identify a few more pixels, then close the Identify Results window.

Figure 2-21

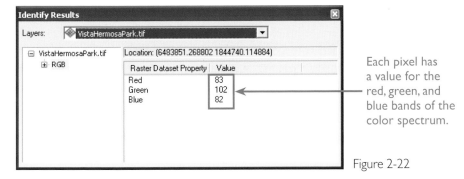

Each pixel has a value for the red, green, and blue bands of the color spectrum.

Figure 2-22

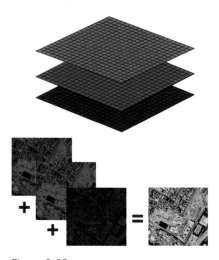

Figure 2-23

Raster data

If you've taken a picture with a digital camera, you're probably familiar with the concept of raster data. A digital image captures a piece of the real world as a two-dimensional array of square cells, or pixels. Each pixel stores a number that corresponds to a specific intensity of light. A picture taken in sunlight is a composite of reflected light in different portions, or bands, of the visible electromagnetic spectrum: blue, green, and red. When these bands are displayed on your computer monitor, they blend to make a color that looks natural (Figure 2-23).

Images are a common form of raster data, but not the only one. Geographic phenomena with a continuous spatial distribution, like elevation and temperature, can also be represented by raster datasets, in which each cell stores a measured or estimated value.

A raster dataset composed of a single matrix of cell values is a single-band raster. One example is a digital elevation model (DEM), where each cell stores a single elevation value. Another example is a grayscale image. True color images, and most satellite images, are multi band, meaning that the raster dataset is composed of multiple spatially coincident cell matrices.

10) Preview census units. In Row 7 of the requirements table we decided to use census units as a proxy for neighborhoods.

Ⓐ In the Catalog Tree, under the census folder, click *tracts.shp* to preview its geography (Figure 2-24).

The data covers Los Angeles County. The tracts to the north are a lot bigger than the ones to the south. That's because census tracts are designed to have a fairly consistent population range, and the northern part of the county, with mountains and desert, is less densely populated.

Ⓑ Click the Zoom In tool ⊕ and zoom in somewhere on the southern part of the county.

Ⓒ In the Catalog Tree, click *block_groups.shp*.

Ⓓ Now click *block_centroids.shp*. If necessary, zoom in farther to see individual points.

Tracts are subdivided into block groups, and block groups are subdivided into blocks. A "block centroid" is a block represented spatially as a point rather than a polygon. (That's not so strange— we've seen the same thing with cities.) You can learn more about census units in the topic *Fundamentals of U.S. Census Geography*.

Ⓔ Click the Full Extent button 🌐.

Figure 2-24

Fundamentals of U.S. Census geography

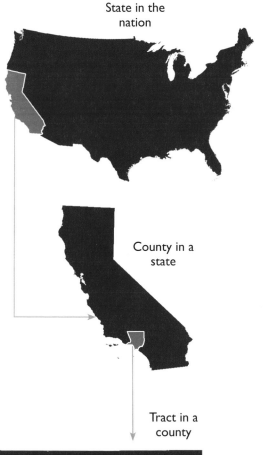

State in the
nation

County in a
state

Tract in a
county

The US Census Bureau reports data by various geographic units. The top-to-bottom relationship shown here represents containment: the nation contains states, states contain counties, counties contain tracts, tracts contain block groups, and block groups contain blocks. Many other non-nesting reporting units are not shown.

Census tracts are relatively small subdivisions of a county. They typically have between 1,500 and 8,000 inhabitants and vary in size according to population density. They are designed to be fairly homogeneous with respect to demographic and economic conditions.

A block group is a cluster of blocks within a tract. A block group typically has between 600 and 3,000 inhabitants.

A census block (most often an ordinary city block) is an area bounded by visible features, like streets or railroad tracks, or by invisible boundaries, like city limits. A block centroid is a census block represented as a point rather than a polygon. A centroid is located in the geographic center of the block it represents and has the attributes of that block.

The Census Bureau conducts a new census every ten years. The latest one was conducted in 2010. Professional demographers estimate values for the intervening years. Updates to census data profoundly affect land-use planning and many other political processes.

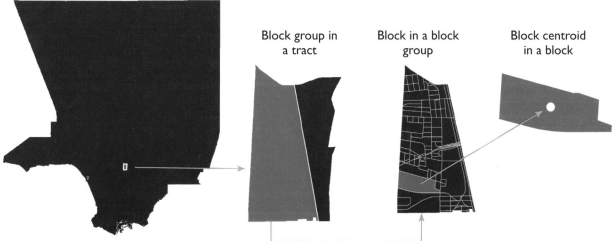

Block group in
a tract

Block in a block
group

Block centroid
in a block

Either block groups or tracts will satisfy our spatial data requirement for a neighborhood. Because of their point geometry, block centroids won't.

Thus far, we've confirmed that we have the spatial and attribute data listed in Rows 1–7 of the data requirements table. In three cases (rivers, parks, census units), we'll have to choose between feature classes. We'll tackle that problem in Lesson 3. We still have more data requirements to consider and more data to preview. We also need to review our requirements for specificity. We'll do this in the next exercise.

F If you're going on to the next exercise, leave ArcCatalog and the data requirements table open. Otherwise, from the ArcCatalog Main menu, choose File → Exit. Save and close the data requirements table.

Exercise 2c: Reframe the problem statement

Some of the city council's guidelines were specific and measurable:

- On a vacant land parcel one acre or larger
- Within the LA city limits

Others were vague:

- As close as possible to the LA River. (Is there a maximum allowed distance from the river? If so, what is it?)
- Not in the vicinity of an existing park. (How close is "in the vicinity"?)
- In a densely populated neighborhood with lots of children. (How densely populated? How many children?)
- In a low-income neighborhood. (How is "low-income" defined?)
- Serving as many people as possible. (How big an area does a park "serve"?)

We can't do the analysis until we eliminate the vagueness.

1)   Define proximity to the LA River. Unless we set a maximum distance limit, every vacant parcel in Los Angeles becomes a potential park candidate. That's absurd on its face and could make us waste a lot of data processing time. We'll set three quarters of a mile as an arbitrary outer limit. That stretches the idea of proximity a little bit, but it's just a cutoff point. Hopefully, we'll find some good locations that are closer than that.

A If necessary, start ArcCatalog. Under C:\UGIS\ParkSite, expand the SourceData folder and its subfolders.

B If necessary, use Windows Explorer to open the data requirements table from the MapsAndMore folder.

By default, you won't see the data requirements table in ArcCatalog. (If you want, you can make *.doc* and other file types visible by choosing ArcCatalog Options from the Customize menu, clicking the File Types tab, and specifying a new file type.)

C In Row 5 of the data requirements table, in the DEFINED AS column, enter `<= 0.75 miles` (Figure 2-25).

Figure 2-25

#	REQUIREMENT	DEFINED AS	SPATIAL DATA	ATTRIBUTE
1	land parcel		parcels	
2	vacant			land use
3	one or more acres			area
4	within city limits		cities	
5	near LA river	<= 0.75 miles	rivers	
6	away from parks		parks	
7	in a neighborhood		census unit	
8	densely populated			population d
9	lots of kids			age
10	low income			income

"<=" means "less than or equal to"

2) Define "not in the vicinity of a park." What minimum distance should a candidate site have to be from existing parks? In open-space planning, 0.25 miles is often used to define a convenient walking distance. (That's typically about a five-minute walk.) Following that standard, we'll say that a site is not in the vicinity of an existing park if the nearest park is at least 0.25 miles away.

A In Row 6 of the requirements table, in the DEFINED AS column, enter `>0.25 miles`.

This measure is a simplification because it's based on straight-line distance. See Appendix A, *Analysis issues,* for further discussion.

3) Define a "densely populated" neighborhood. As we make the rest of the requirements concrete, we'll also make sure that we have the appropriate data.

A In ArcCatalog, in the Catalog Tree, under the census folder, click *tracts.shp*.

B Preview its table and scroll across the attributes.

As shown in Figure 2-26, there are two population density attributes: POP00_SQMI and POP07_SQMI (population per square mile for 2000 and 2007). The 2000 numbers are Census Bureau data; the 2007 numbers are projections by Esri demographers.

Figure 2-26

FIPS	POP2000	POP2007	POP00_SQMI	POP07_SQMI
06037101110	4500	4801	10227.3	10911.4
06037101120	3280	3640	658.6	730.9
06037101210	6066	6303	24264	25212
06037101220	3028	3269	11214.8	12107.4
06037101300	3974	4080	4096.9	4206.2
06037101400	3760	3862	1613.7	1657.5
06037102101	3387	3452	454	462.7
06037102102	6739	7067	4522.8	4743
06037103101	2610	2693	1641.5	1693.7
06037103102	4462	4729	9295.8	9852.1
06037103200	5484	5615	1979.8	2027.1
06037103300	3739	3913	1027.2	1075
06037103400	6209	6524	4005.8	4209
06037104103	3840	3930	10971.4	11228.6
06037104104	3507	4314	5845	7190

Contents | Preview | Description

Preview: Table

1 (of 2054)

c Preview the *block_groups.shp* table and scroll across its attributes.

There's a total population attribute (TOTPOP_CY), but not one for population density. Given an AREA attribute, we can calculate density: it's just population divided by area. And even though this table doesn't have an AREA attribute either, we can make one for it in ArcMap. So both *tracts* and *block_groups* have the needed attribute: one explicitly, the other implicitly.

We still need a definition of "densely populated." To keep it simple, let's call a neighborhood densely populated if its value exceeds that of Los Angeles as a whole. The population density of Los Angeles as of the year 2000, was 7,876.8 people per square mile. Rounding up to an even number, we'll use 8,000 people per square mile as our threshold.

D In Row 8 of the data requirements table, in the DEFINED AS column, enter **>= 8,000 per sq mi**.

Threshold values

To find the average population density of Los Angeles, as well as other threshold values for the analysis, we used online U.S. Census Bureau data, especially the American FactFinder website at http://factfinder.census.gov.

4) Define "lots of children." Again, we'll look for attributes and then set a threshold.

A In ArcCatalog, scroll in the *block_groups* table and locate the POP18UP_CY attribute (Figure 2-27).

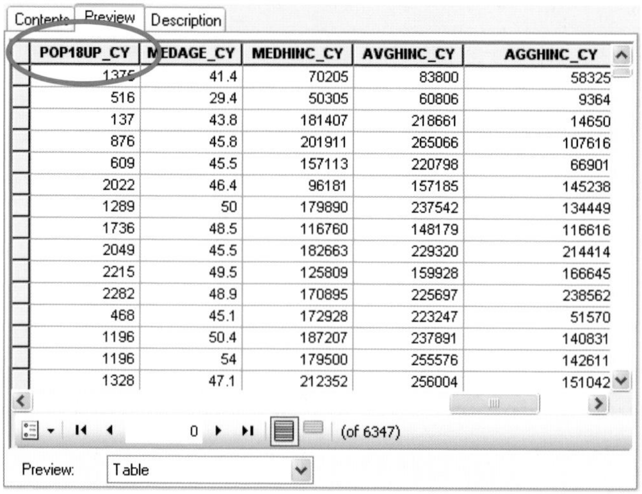

POP18UP_CY	MEDAGE_CY	MEDHINC_CY	AVGHINC_CY	AGGHINC_CY
1375	41.4	70205	83800	58325
516	29.4	50305	60806	9364
137	43.8	181407	218661	14650
876	45.8	201911	265066	107616
609	45.5	157113	220798	66901
2022	46.4	96181	157185	145238
1289	50	179890	237542	134449
1736	48.5	116760	148179	116616
2049	45.5	182663	229320	214414
2215	49.5	125809	159928	166645
2282	48.9	170895	225697	238562
468	45.1	172928	223247	51570
1196	50.4	187207	237891	140831
1196	54	179500	255576	142611
1328	47.1	212352	256004	151042

Figure 2-27

This is the population 18 years old or older for the current year. If we define a child as a person under 18 (which seems reasonable), then we can subtract the values in this field from those in TOTPOP_CY to get the number of children.

❸ Preview the *tracts* table and scroll across its attributes.

Here, the population is segmented into age ranges (Figure 2-28). We can find the number of children by adding the values in the AGE_UNDER5 and AGE_5_17 fields. So again, we can obtain the required attribute from either census table.

Contents	Preview	Description				
AGE_UNDER5	**AGE_5_17**	**AGE_18_21**	**AGE_22_29**	**AGE_30_39**	**AGE_40_**	
341	853	221	484	757		
200	572	192	190	496		
607	1274	362	918	1220		
226	561	163	306	583		
237	596	145	249	581		
227	647	154	299	590		
185	639	161	236	460		
367	1017	422	708	1023	1	
170	514	96	164	462		
279	907	205	388	718		
356	884	232	522	830		
177	597	141	200	550		
414	1072	242	585	1017	1	
296	982	270	430	559		
269	785	230	421	499		

Preview: Table

1 ▶ ▶| (of 2054)

Adding values in these two fields gives the total number of children.

Figure 2-28

Because neighborhoods vary in size and population, we can make more valid comparisons if we base our threshold on a ratio rather than on an absolute number. (We can derive the percentage of children from our data with some simple arithmetic.) In Los Angeles as a whole, 75.2 percent of the population is 18 or older, meaning that children make up 25 percent of the population. We'll therefore define a neighborhood as having "lots of children" if it meets or exceeds this value.

❸ In Row 9 of the requirements table, in the DEFINED AS column, enter **>= 25%**.

❹ In the ATTRIBUTE DATA column, replace the entry "age" with **age under 18**.

5) **Define "low income."** Let's preview our income attributes.

❶ In ArcCatalog, in the *tracts* table, scroll across the fields.

There don't appear to be any income attributes.

❷ Preview the *block_groups* table and scroll all the way to the right.

Metadata

Metadata is a description of what you know about a dataset. It serves two important purposes. First, it builds user confidence in the data by explaining things like how, when, and by whom the data was created. Without this confidence, no one would risk using the data for serious work. Second, it makes the data searchable. Metadata includes keywords that identify thematic, geographic, and temporal properties of the data (for example, "rivers," "Los Angeles," "2010"). Keywords make it possible to find datasets among large inventories of spatial data.

ArcGIS supports basic and complete metadata styles. A "style" here means both a standard, defining how much documentation is kept, and a framework for organizing and presenting the documentation. The most basic, and default, metadata style is the Item Description. It supports keywords, descriptions of the dataset's contents and intended use, acknowledgment of the data creator, and conditions for permitted use of the data. Anyone who creates and shares data should provide this much information.

The other styles are for complete metadata, as maintained by government agencies, commercial data vendors, and many large enterprises. The complete metadata styles are fairly similar: the differences often amount to things like whether a certain piece of information is optional or required. You may never have to create complete metadata, but you will certainly use datasets that have complete metadata. You can still use the Item Description style with these datasets—it just means that all the extra information will be filtered out. To see the complete metadata, you have to change the metadata style. For data created in the United States or Canada, use the North American Profile of ISO 19115 2003. (ISO 19115 2003 is a metadata standard developed by the International Organization for Standardization, a network of the national standards institutes of 163 countries.)

When you use one of the complete metadata styles, the item description will expand to show ArcGIS Metadata and FGDC Metadata headings. For the most part, these two headings organize and present the same metadata

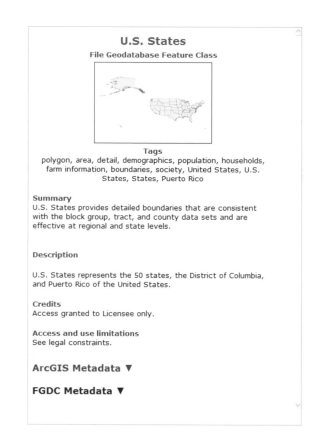

U.S. States

File Geodatabase Feature Class

Tags
polygon, area, detail, demographics, population, households, farm information, boundaries, society, United States, U.S. States, States, Puerto Rico

Summary
U.S. States provides detailed boundaries that are consistent with the block group, tract, and county data sets and are effective at regional and state levels.

Description

U.S. States represents the 50 states, the District of Columbia, and Puerto Rico of the United States.

Credits
Access granted to Licensee only.

Access and use limitations
See legal constraints.

ArcGIS Metadata ▼

FGDC Metadata ▼

in different formats. They aren't exactly the same, however. ArcGIS metadata is a new format at ArcGIS 10. It includes a richer set of "automatic" metadata elements; that is, elements maintained by the software itself. Examples of automatic metadata include the dataset's coordinate system, attribute names and data types (string, numeric, and so on), and a history of data processing operations on the dataset. These examples are found in both ArcGIS and FGDC metadata.

When you create new metadata or update old metadata in ArcGIS 10, it becomes ArcGIS metadata: the rich set of automatic elements is generated for it. Metadata created outside ArcGIS can be upgraded to ArcGIS metadata with the Upgrade Metadata tool.

The last four attributes might be income measures, but we can't be sure without decoding the cryptic field names. To do this, we'll look at the metadata—not the item descriptions we looked at before, but the full data documentation.

ⓒ From the main menu, choose Customize → ArcCatalog Options. In the dialog box, click the Metadata tab.

ⓓ Change the metadata style from Item Description to North American Profile of ISO19115 2003 (Figure 2-29). Click OK.

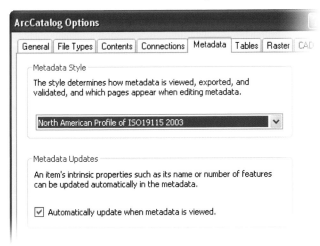

Figure 2-29

See the topic *Metadata* for more information.

ⓔ Click the Description tab and scroll to the bottom of the display window.

At the bottom of the window are headings for ArcGIS Metadata and FGDC Metadata that weren't there before.

ⓕ Click the ArcGIS Metadata heading to expand it.

More headings categorize the metadata.

ⓖ Near the bottom, click the ESRI Fields and Subtypes heading. Scroll down through the metadata.

Each field is described by several pieces of information. Asterisked information (*) was automatically created by ArcGIS. Non-asterisked information was entered by a person.

ⓗ Scroll down to the last field, PCI_CY, and read the field description.

The field is per capita income for 2008. (The values are estimates based on 2000 census data.)

ⓘ Read the descriptions for the three fields above PCI_CY.

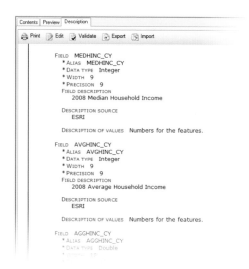

Figure 2-30

The attributes are aggregate (total) household income, average household income, and median household income (Figure 2-30).

We don't want to use aggregate income because it's not standardized: some block groups have more people than others. The other three are all possible. Median income is a statistical midpoint: it marks the value that half the households are above and half are below. We'll adopt this measure on the grounds that it's less sensitive to extreme values. (A millionaire in a low-income neighborhood might significantly change the average income, but not the median income.)

The median household income for Los Angeles is $48,610. Rounding off, we'll call a neighborhood "low income" if the median household income is $50,000 or less.

J In Row 10 of the requirements table, in the DEFINED AS column, enter **<= $50,000**.

K In the ATTRIBUTE DATA column, replace "income" with **median hh income**.

6) Define "serving the most people." Finally, we want to know which potential site serves the most people. Anyone can come to a park, so we need to count all the people nearby, regardless of demographic profile. The attribute we need for this is total population, which we have in both the *tracts* and *block_groups* feature classes.

We'll treat this guideline as a preference. If half a dozen sites meet our other requirements, we prefer those serving more people to those serving fewer. Eventually, this preference may have to be subjectively weighed against others. For example, which is better: a park closer to the river that serves fewer people, or a park farther from the river that serves more people? (How much closer? How many more people?)

To define the size of the area served by a park, we'll apply the standard of easy walking distance discussed earlier, and say that a park serves anyone who lives within a quarter mile of it. "Serving the most people" therefore means having the largest population within a 0.25 mile radius.

A In ArcCatalog, preview the *block_centroids* table and scroll across its attributes.

This table also has a total population attribute (POP2000). We can't use block centroids as our spatial data for neighborhoods—we need polygons—but we could use them to sum population. Given the boundary of a park service area, ArcMap could count the block points that fall within it and add their population values.

B In Row 11 of the requirements table, in the DEFINED AS column, enter **<= 0.25 miles**.

Your requirements table should look like Figure 2-31.

#	REQUIREMENT	DEFINED AS	SPATIAL DATA	ATTRIBUTE DATA	DATA SET	PREPARATION
1	land parcel		parcels			
2	vacant			land use		
3	one or more acres			area		
4	within city limits		cities			
5	near LA river	<= 0.75 miles	rivers			
6	away from parks	> 0.25 miles	parks			
7	in a neighborhood		census unit			
8	densely populated	>= 8,000 per sq mi		population density		
9	lots of kids	>= 25%		age under 18		
10	low income	<= $50,000		median hh income		
11	serving the most people	<= 0.25 miles		population		
12	final map		political boundaries			
13	final map		roads			
14	final map		relief			
15	final map		imagery			

Figure 2-31

C From the ArcCatalog main menu, choose Customize → ArcCatalog Options.

D On the Metadata tab, change the style back to Item Description. Click OK.

E From the ArcCatalog main menu, choose File → Exit.

F Save and close the data requirements table.

We can now state the analysis problem in measurable terms, which means that we can solve it with GIS tools. Someone might take issue with our definitions, but that's fine—we'll always be happy to improve our methodology. For now, we can state our project analysis requirements as follows:

We want to locate a site for a new park

- On a land parcel
- That is vacant
- One or more acres in size
- Within the LA city limits
- Within 0.75 miles of the LA River (preferring closer sites)
- More than 0.25 miles from the nearest park
- In a neighborhood with a population density of 8,000 or more people per square mile
- Where at least 25 percent of the population is under 18 years old
- Where median household income is $50,000 or less
- With preference given to sites that serve more people within a 0.25 mile radius

We've formalized the data requirements in a table and confirmed that the data is available. In some cases, there are multiple datasets describing the same features and attributes. In the next lesson, we'll compare these datasets and choose which ones to use in the analysis.

Lesson

3 Choose the data

CHOOSING THE DATASETS TO USE

in the analysis is our first goal in this lesson. We came across a few situations in Lesson 2 (with rivers, parks, and census units) in which different datasets stored similar features and attributes. How do we choose one dataset over another? Here are some considerations:

- One may represent features with a more suitable geometry type for your needs (think of cities as polygons versus cities as points).
- Among datasets of the same geometry type, one may represent features with a more appropriate degree of detail.
- One dataset may be of more recent vintage than another. Or it may be more complete, including features that the other lacks.
- One dataset may be more accurate than another when viewed against imagery or other basemaps. Or it may have been created by a more authoritative source. Or it may conform better to another dataset you've already decided to use.
- One dataset may need less preliminary processing of spatial or attribute data to make it ready for analysis.

Our second goal is to choose a coordinate system for the analysis. In reality, that decision is often determined by prevailing standards in your organization, or by more or less default best choices for a given study area. For example, it makes sense for us to use the State Plane Coordinate System of 1983 because it is specifically designed to minimize spatial distortion for more than a hundred local areas (like ours) within the United States. While the choice of a coordinate system bears directly on our project, Exercise 3b also contains background material about coordinate systems (what they are, why they matter, how they're managed) that is important to know for any GIS work.

Lesson Three roadmap

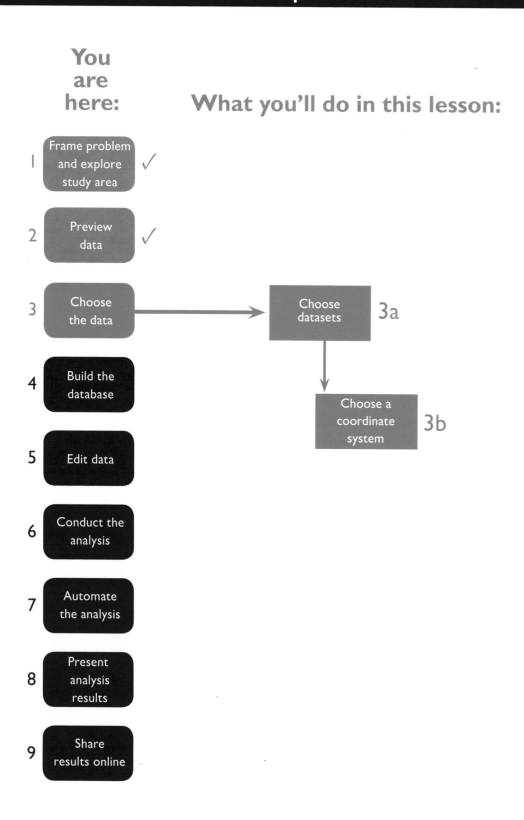

You are here:

What you'll do in this lesson:

1 Frame problem and explore study area ✓

2 Preview data ✓

3 Choose the data → Choose datasets 3a

→ Choose a coordinate system 3b

4 Build the database

5 Edit data

6 Conduct the analysis

7 Automate the analysis

8 Present analysis results

9 Share results online

Exercise 3a: Choose datasets

In this exercise, we'll add layers to ArcMap, compare them, and decide which datasets to use in the analysis. We'll add this information to the data requirements table we used in the last lesson.

1) Start ArcMap. We'll work in ArcMap in this lesson and access data from the Catalog window.

Ⓐ Click the Windows Start button, then choose All Programs → ArcGIS→ ArcMap 10.

▷ If you have a shortcut on your desktop, you can double-click the ArcMap 10 icon **Q** instead.

Ⓑ In the Getting Started dialog box, under New Maps, click My Templates to highlight it (Figure 3-1). Click Blank Map, then click OK.

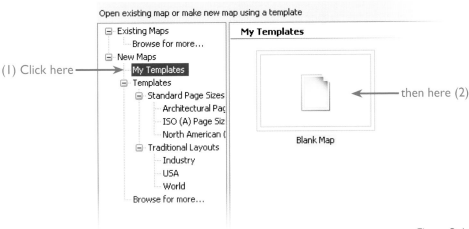

Figure 3-1

Ⓒ Make sure the Catalog window is either open or visible as a tab on the right side of the ArcMap window.

▷ If necessary, on the Standard toolbar, click the Catalog window button 🗊.

2) Choose the parcel data. Parcels are the individual property boundaries from which our candidate park sites will eventually be selected.

Ⓐ In the Catalog window, click the plus sign next to Folder Connections to expand it.

Ⓑ Expand C:\UGIS.

Ⓒ Expand the ParkSite, SourceData, and City of LA workspaces.

In Lesson 2, we previewed the *Parcels* shapefile and the *Vacant Parcels* table in the City of LA folder. We don't have any other parcel data, so there's no real choice to make here.

D Using Windows Explorer, open the data requirements table from C:\UGIS\ParkSite\MapsAndMore.

> ▷ If you didn't complete Lesson 2, you can download the lesson results from the Understanding GIS Resource Center.

E In Row 1 of the table, in the DATASET column, enter **Parcels**.

F In Row 2, under DATASET, enter **Vacant Parcels**.

We need to join the *Vacant Parcels* table to the *Parcels* attribute table to see vacant parcels on the map. We'll do this table join—along with all our other data preparation—in Lesson 4, but we'll make a note of it here.

G In Row 2 of the requirements table, in the PREPARATION column, enter **join table to Parcels**.

We also need to calculate parcel acreage to satisfy our third requirement.

H In Row 3, under DATASET, enter **Parcels**. Under PREPARATION, enter **calculate area** (Figure 3-2).

Figure 3-2

#	REQUIREMENT	DEFINED AS	SPATIAL DATA	ATTRIBUTE DATA	DATASET	PREPARATION
1	land parcel		parcels		Parcels	
2	vacant			land use	Vacant Parcels	join table to Parcels
3	one or more acres			area	Parcels	calculate area
4	within city limits		cities			
5	near LA river	<= 0.75 miles	rivers			
6	away from parks	> 0.25 miles	parks			
7	in a neighborhood		census unit			

3) Choose the city limits data. The requirement in Row 4 is that the new park be within the city limits. We assumed this meant that we'd need spatial data representing the boundary of Los Angeles (which we have), but actually, there's an even simpler solution.

A In ArcMap, in the Catalog window, under City of LA, right-click *Vacant Parcels.dbf* and choose Item Description.

The Item Description window opens. Its Description and Preview tabs give you access to the same information you got from ArcCatalog in the last lesson.

B In the Item Description window, click the Preview tab.

Figure 3-3

C Scroll across the attributes and make sure you can see the ZONECITY field (Figure 3-3).

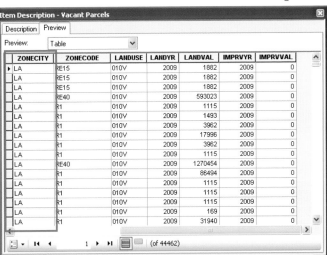

This field tells you which city each vacant parcel is recorded in.

D Make sure you're scrolled to the top of the table. Right-click the ZONECITY field heading and choose Sort Ascending.

The first record is in LA.

E Right-click the ZONECITY field heading again and choose Sort Descending.

The table doesn't change, which tells you that every record in the field has the same value. This means that all our vacant parcels—and consequently, all our potential park sites—are prequalified as belonging to Los Angeles. It turns out that we don't need a spatial boundary after all to guarantee that a parcel lies within the city limits.

You can resize the window by dragging any side or corner.

F Close the Item Description window.

G In Row 4 of the requirements table, under DATASET, enter **Vacant Parcels**.

H Also in Row 4, under SPATIAL DATA, delete "cities." Under ATTRIBUTE DATA, enter **city name**.

4) Add and symbolize the river data. To analyze the distance of parcels to the river, we need spatial data representing the Los Angeles River. We can use either the *River* feature class in the ESRI geodatabase or the *LARiver* shapefile. We'll compare these two datasets against an imagery basemap.

Figure 3-4

A In ArcMap, on the Standard toolbar, click the Add Data drop-down arrow and choose Add Basemap.

B In the Add Basemap dialog box, click Imagery and click Add.

By default, the *Basemap* layer is zoomed to the extent of the whole world.

C In the Catalog window, expand ESRI.gdb and expand Hydro.

D Drag and drop *River* onto the map display.

E In the Table of Contents, right-click the *River* layer and choose Zoom To Layer (Figure 3-4).

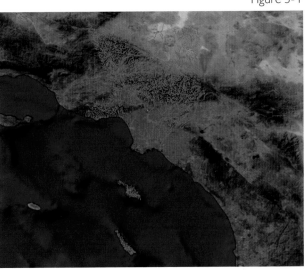

Layers usually draw in randomly-selected colors. You may not have noticed, but this layer has drawn in exactly the same shade of blue both times you've added it to ArcMap. Let's see why.

F In the Table of Contents, click the symbol patch for the *River* layer to open the Symbol Selector (Figure 3-5).

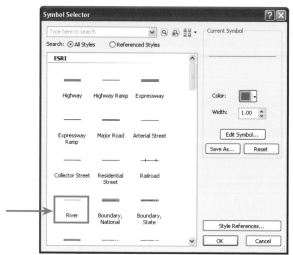

When a feature class name matches a symbol name, that symbol is assigned automatically.

Figure 3-5

In the Symbol Selector, notice the River symbol in the scrolling box. Because the feature class name (*River*) matches the name of a predefined symbol, the layer is symbolized automatically.

G Close the Symbol Selector.

Symbols and styles

ArcGIS has thousands of predefined symbols for map features, organized into "styles" on a mostly thematic basis: the Caves style, the Crime Analysis style, and so on. By default, symbols from two styles are loaded in the Symbol Selector. One is the generic Esri style, which contains common map symbols; the other is your personal style (which is empty unless you add symbols to it). The set of available symbols changes according to whether you're symbolizing point, line, or polygon features.

To load additional styles, click the Style References button on the Symbol Selector, then check the styles you want. Symbols from these styles will be added at the bottom of the Symbol Selector. You can also search for symbols by typing a keyword in the Search box at the top of the Symbol Selector.

Styles contain not only feature symbols, but also labels, color ramps, north arrows, scale bars, and many other map elements. To view or edit the contents of a style, choose Style Manager from the Customize menu.

The only river we want to see on the map is the LA River. In Lesson 1, we solved this problem with a definition query. Another approach is to symbolize the Los Angeles River and no other features.

H Open the layer properties for the *River* layer and click the Symbology tab.

I In the Show box, click Categories.

You symbolize by category when the attribute of interest is a name, a description, or a number that isn't a quantity, like a code or ranking. The default Unique Values method assigns a different symbol to each unique value in the specified field.

J Make sure the Value Field drop-down list is set to NAME. Click the Add Values button (Figure 3-6).

Figure 3-6

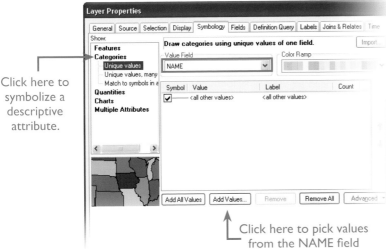

Click here to symbolize a descriptive attribute.

Click here to pick values from the NAME field

Figure 3-7

K In the Add Values dialog box, click Complete List.

L Scroll down to "Los Angeles River." (The list is alphabetical.) Click on it to highlight it (Figure 3-7), then click OK.

On the Symbology tab, "Los Angeles River" is added and assigned a symbol from the current color ramp. By default, other features in the layer will get the symbol next to <all other values>. We don't want to see those features, however.

M Uncheck the <all other values> check box.

N Double-click the symbol patch next to Los Angeles River to open the Symbol Selector.

O Change the color to Apatite Blue (Figure 3-8). Change the line width to 2. Click OK on the Symbol Selector.

Figure 3-8

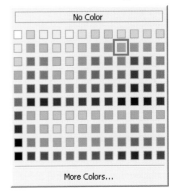

P Compare your settings to Figure 3-9, then click OK on the Layer Properties dialog box.

Uncheck this box ⟶

Symbol	Value	Label	Count
☐	<all other values>	<all other values>	3574
	<Heading>	**NAME**	**17**
——	Los Angeles River	Los Angeles River	17

Figure 3-9

On the map, only the Los Angeles River is symbolized (Figure 3-10).

Q In the Catalog window, under City of LA, drag and drop *LARiver.shp* onto the map.

You should get a Geographic Coordinate Systems Warning (Figure 3-11). If the meaning of this warning is unclear to you—as it probably will be unless you know how ArcGIS manages coordinate systems—be patient until we get to Exercise 3b.

Figure 3-10

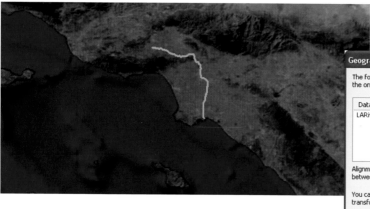

Figure 3-11

R Click Close on the Geographic Coordinate Systems Warning dialog box.

The *LARiver* layer is added to the map.

S In the Table of Contents, click the *LARiver* symbol patch to open the Symbol Selector.

T Change the color to Cretean Blue (Figure 3-12) and the width to 2. Click OK.

Figure 3-12

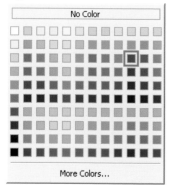

5) **Choose the river data.** Even at this relatively small (zoomed-out) scale, it's obvious that the two representations of the river have different spatial extents. One goes to the sea while the other stops abruptly halfway there. Let's find out why.

Ⓐ In the Catalog window, expand the MapsAndMore folder.

Ⓑ Drag and drop *Los Angeles.lyr* on the map.

▷ If you didn't create this file in Lesson 1, you can download the Lesson 1 results from the Understanding GIS Resource Center.

Ⓒ In the Table of Contents, right-click the *Los Angeles* layer and choose Zoom To Layer (**Figure 3-13**).

This is the layer file (the file of saved layer properties) that you made in Lesson 1. When you add it to the map, its layer properties are already set.

Ⓓ Open the layer properties for the *Los Angeles* layer.

Ⓔ On the Layer Properties dialog box, click the Symbology tab, then the Definition Query tab.

As reflected on the map, the symbol is a hollow yellow outline and there is a definition query on Los Angeles.

Ⓕ Click the Source tab (**Figure 3-14**).

Figure 3-13

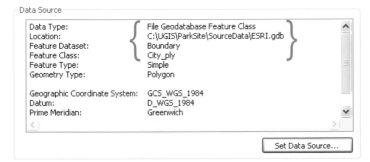

Figure 3-14

A layer file isn't a feature class—it doesn't store feature coordinates and attributes. Every layer in a map document must point to a feature class on disk, and the same goes for a layer file. This layer file points to the *City_ply* feature class.

Ⓖ Close the Layer Properties dialog box.

We can see the likely explanation for the different extents of the two layers: in the *LARiver* layer, the river stops at the city boundary. The City of Los Angeles, which provided the source data for this layer, doesn't need to maintain features beyond its own jurisdiction. (It's interesting that even a "natural" feature like a river may be defined in part by administrative or political considerations.)

H Zoom in ⊕ on the source (the northwestern end) of the river. Set the map scale to 1:24,000.

I Pan ✋ along the river and visually compare the two representations. Go all the way to where the *LARiver* layer stops at the city boundary.

The two representations of the river aren't identical, but they're pretty close. At very large scales, like 1:2,500, the river in *LARiver* is noticeably more sinuous and conforms more closely to the imagery, but for our analysis, either dataset is perfectly adequate. We'll use the *LARiver* dataset, which is already clipped to our area of interest.

J In Row 5 of the table, under DATASET, enter **LARiver**.

When we previewed this dataset in Lesson 2, we saw that it was composed of 265 features. We can combine these into a single feature with a data processing operation called Dissolve. That will help get the data ready for Lesson 6, where we'll need to create a 3/4 mile buffer (proximity zone) around the river. We want one zone, not 265.

K In Row 5, under PREPARATION, enter **dissolve** (Figure 3-15).

Figure 3-15

#	REQUIREMENT	DEFINED AS	SPATIAL DATA	ATTRIBUTE DATA	DATASET	PREPARATION
1	land parcel		parcels		Parcels	
2	vacant			land use	Vacant Parcels	join table to Parcels
3	one or more acres			area	Parcels	calculate area
4	within city limits			city name	Vacant Parcels	
5	near LA river	<= 0.75 miles	rivers		LARiver	dissolve
6	away from parks	> 0.25 miles	parks			
7	in a neighborhood		census unit			
8	densely populated			population density		

6) Save the map document. This is a good time to save the map document. Let's take a minute to understand how ArcMap encourages us to organize maps and data.

A In ArcMap, open the Catalog window.

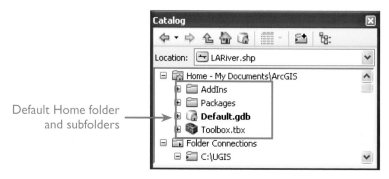

Default Home folder and subfolders

Figure 3-16

At the top of the window is a folder called "Home" (Figure 3-16). Every map document has a home folder, which is simply the folder in which it is saved. In a new ArcMap session, before a map document has been saved, the designated home folder is MyDocuments\ArcGIS.

B On the Standard toolbar, click the Save button ![save icon].

C Navigate, if necessary, to the MapsAndMore folder. Name the map document **Lesson3a** and click Save.

In the Catalog window, the home folder changes to ParkSite\MapsAndMore (Figure 3-17). The name of the open map document is boldfaced.

After saving

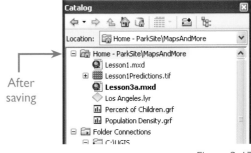

Figure 3-17

Home folders and default geodatabases

The home folder is intended to help you keep map documents, datasets, and related files in the same place. This keeps the path relationship between layers and their data sources simple and easy to manage. It also simplifies the creation of map packages, which are bundles of maps and data that you can share (see Lesson 9). Most nondataset files that you save from a map document, like layer files, graphs, and bookmarks, are directed to the home folder.

In addition to a home folder, every map document has a default geodatabase. This is the location to which data processing outputs are directed by default. Unless otherwise specified, the default geodatabase (which is actually named "Default") is located in MyDocuments\ArcGIS. You can choose a different default geodatabase when you start a new map document. You can also change the default geodatabase during a session by right-clicking the geodatabase you want in the Catalog window and choosing Make Default Geodatabase.

7) **Add and symbolize the park data.** To analyze the location of candidate sites with respect to existing parks, we need spatial data representing parks. We'll compare the *Parkland* dataset we used in Lesson 1 to the *Parks* dataset we previewed in Lesson 2.

A In the Table of Contents, turn off the *River* layer and click its minus sign to collapse it.

▷ Leave the *LARiver* layer turned on.

B In the Catalog window, under ESRI.gdb, expand Landmark.

C Drag and drop *Parkland* onto the map.

D From the City of LA folder, drag and drop *Parks.shp* onto the map.

▷ Close the Geographic Coordinate Systems Warning if you get it.

Figure 3-18

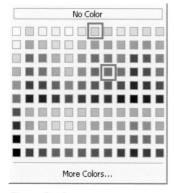

Figure 3-19

The order of layers in the Table of Contents should look like Figure 3-18. Your symbology for the parks layers will be different.

E In the Table of Contents, right-click the symbol patch for the *Parks* layer to open the color palette. Change the fill color to Chrysophase (Figure 3-19).

F Right-click the *Parkland* symbol patch and change its fill color to Tzavorite Green.

G Open the layer properties for the *Parks* layer and click the Display tab.

H In the Transparent % box, type **50**. Click OK.

The symbology makes it clear where the two layers agree and disagree on the representation of features (Figure 3-20).

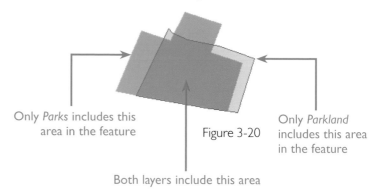

Only *Parks* includes this area in the feature

Both layers include this area

Figure 3-20

Only *Parkland* includes this area in the feature

8) Examine a park in two layers. Let's start with a close look at an example park. Your map should still be zoomed in to where the river crosses the city limits. The scale should still be 1:24,000.

A On the Tools toolbar, click the Measure tool 📐. At the top of the dialog box, make sure the Measure Line tool ～ is selected.

B Click the Choose Units button ▾ , point to Distance, and choose Miles, if necessary (Figure 3-21).

C Move the Measure dialog box away from the river, if necessary.

We're going to follow the river north about two and three-quarters miles.

D Click at the end of the river to start the measurement. Move the mouse pointer north along the river (you don't have to drag).

E Watch the Length value change in the Measure dialog box as you go.

F When you reach the top of the map display—assuming your measurement is still less than 2.75 miles—press and hold the letter C on your keyboard.

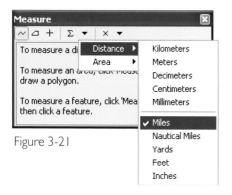

Figure 3-21

The mouse pointer changes to the Pan icon 🖑.

G Pan north, then release the C key to continue the measurement.

H When the length of the measurement reaches about 2.75 miles, double-click to end the measurement.

Shortcut keys

You can use these keys as navigation shortcuts:

  Z to zoom in; X to zoom out; C to pan

  B for zoom/pan (click and drag to zoom; right-click and drag to pan)

  Q to roam (the map moves continuously)

There are shortcuts for many other ArcGIS operations. In the ArcGIS Desktop Help, search for the keyword "shortcuts" to learn more.

Half a mile due east of the end of your measurement is a park.

I Identify 🛈 the park.

Its name is Pecan Playground (Figure 3-22). The attributes in the Identify window come from the *Parks* layer, not from *Parkland,* because of the layer order in the Table of Contents.

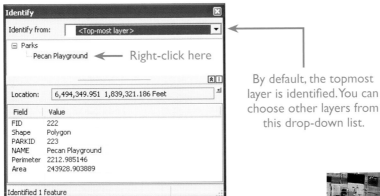

Right-click here

By default, the topmost layer is identified. You can choose other layers from this drop-down list.

Figure 3-22

Figure 3-23

J In the top pane of the Identify window, right-click the name Pecan Playground. On the context menu, choose Zoom To.

The map zooms in on the park (Figure 3-23).

K Close the Identify window and the Measure dialog box.

L Turn the *Parkland* layer off and on a few times.

Neither feature conforms to the image: both encroach, for example, on the surrounding streets. The *Parks* feature seems better, however, because it includes the swimming pool and the play area at the north end of the block.

Figure 3-24

Data Frame Tools toolbar

We'll come back to this place in Lesson 5 to do some spatial editing. We noted in Lesson 1 that a tool called My Places is a good way to store and manage locations independently of particular map documents. (Within a map document, it may be easier to use spatial bookmarks.) To use My Places, we have to add a toolbar.

Ⓜ From the ArcMap main menu, choose Customize →Toolbars → Data Frame Tools (Figure 3-24).

Ⓝ On the toolbar, click the My Places button 🏳.

Ⓞ On the right side of the My Places dialog box, click Add From and choose Current Extent.

An entry called My Place is added to the box (Figure 3-25).

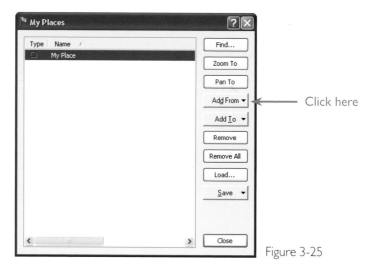

Figure 3-25

Ⓟ Right-click the highlighted entry and choose Rename. Replace the default name with **Pecan Playground** and press Enter.

In Lesson 1, we saved our spatial bookmarks to an external file. We'll load these into My Places now.

Ⓠ In the My Places dialog box, click Load. Navigate, if necessary, to the MapsAndMore folder.

Ⓡ Click Lesson 1 Places.dat to highlight it and click Open.

▷ If you don't have this file, you can download the Lesson 1 results from the Understanding GIS Resource Center.

Your My Places dialog box should look like Figure 3-26.

Figure 3-26

❻ In the My Places dialog box, click Dodger Stadium to highlight it, then click Zoom To.

Your view is now centered on Dodger Stadium.

❼ Close the My Places dialog box.

9) **Choose the park data.** We'll follow the river upstream and look at some more parks to help us reach a decision.

Ⓐ In the Table of Contents, drag the *Los Angeles* layer directly underneath the *LARiver* layer.

Ⓑ Set your map scale to **1:36,000**. Start panning northwest along the river's course.

You'll soon come to Griffith Park, the vast park at the river's elbow. Both the river and the city limits run along the park's edge. Notice that to the east (Glendale) and north (Burbank) of the city limits, there are features from the *Parkland* layer, but no corresponding *Parks* features (Figure 3-27). This is another case of jurisdiction. The City of Los Angeles, which supplied the *Parks* dataset, doesn't maintain data for parks in other cities.

Figure 3-27

That raises a question. The new park has to be a quarter mile away from existing parks: does that mean any park in any city or just any park in Los Angeles? We'll interpret it to mean any park in any city, since by and large, there are no residency requirements for park use. On that basis, we should choose the *Parkland* feature class for our analysis. We don't know if it's more or less accurate on the whole (we only examined one case), but it covers a part of our area of interest that the *Parks* feature class doesn't.

Ⓒ In Row 6 of the requirements table, under DATASET, enter **Parkland**.

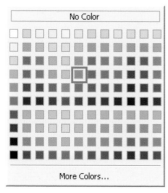

Figure 3-28

10) Add more parks data. In Lesson 2, we previewed the *NewParks* feature class and the *VistaHermosaPark* raster dataset. We should find out if these features are included in the *Parkland* layer.

Ⓐ In ArcMap, in the Catalog window, expand the ParkData folder. Add *NewParks.shp* to the map.

▷ Close the Geographic Coordinate Systems Warning if you get it.

Ⓑ In the Table of Contents, drag the *NewParks* layer directly above the *Parks* layer.

Ⓒ Right-click the symbol patch for the *NewParks* layer and click Macaw Green (Figure 3-28).

Ⓓ Zoom to the *NewParks* layer (Figure 3-29).

▷ How? Right-click the layer in the Table of Contents and choose Zoom To Layer.

Figure 3-29

Ⓔ Turn the *NewParks* layer off and on a couple of times.

You can see parks on the basemap imagery, but they're not represented in the *Parkland* layer. Another data preparation task in Lesson 4 will be to load these two park features into the *Parkland* feature class.

Ⓕ In the Catalog window, under ParkData, add *VistaHermosaPark.tif* to the map.

▷ Close the Geographic Coordinate Systems Warning if you get it.

Ⓖ In the Table of Contents, collapse the layer by clicking its minus sign.

Ⓗ Zoom to the *VistaHermosaPark.tif* layer.

There's no *Parkland* feature here, either, or we'd be looking at it. In Lesson 5, we'll create a park feature in this location.

I Turn the *VistaHermosaPark.tif* layer off and on a couple of times (Figure 3-30).

The basemap imagery, in which the park looks more developed, is probably a little newer.

J Open My Places 🏳 . Click Add From and choose Current Extent.

K Right-click the new entry and choose Rename. Name it **Vista Hermosa Park**. Click Close on the My Places dialog box.

L In Row 6 of the requirements table, under PREPARATION, enter **load data and edit features**.

11) Choose the census data. In Lesson 2, we decided that we would represent neighborhoods with census units. Our two choices are block groups or tracts. Block groups meet all our attribute needs, either directly or indirectly. Because they are smaller than tracts, block groups also portray demographic patterns in more detail.

A In Rows 7 through 10 of the requirements table, under DATASET, enter **block_groups**.

Population density and age under 18 attributes don't exist as such in the *block_groups* table, but we can derive them.

B In Rows 8 and 9, under PREPARATION, enter **calculate**.

The last analytical requirement is to count how many people live within 0.25 miles of each proposed park site. In Lesson 6, we'll do this by drawing quarter-mile rings (buffers) around the proposed sites, counting the census block points in each ring, and summing their populations.

C In Row 11, under DATASET, enter **block_centroids** (Figure 3-31).

Figure 3-30

Figure 3-31

#	REQUIREMENT	DEFINED AS	SPATIAL DATA	ATTRIBUTE DATA	DATASET	PREPARATION
1	land parcel		parcels		Parcels	
2	vacant			land use	Vacant Parcels	join table to Parcels
3	one or more acres			area	Parcels	calculate area
4	within city limits			city name	Vacant Parcels	
5	near LA river	<= 0.75 miles	rivers		LARiver	dissolve
6	away from parks	> 0.25 miles	parks		Parkland	load data and edit features
7	in a neighborhood		census unit		block_groups	
8	densely populated	>= 8,000 per sq mi		population density	block_groups	calculate
9	lots of kids	>= 25%		age under 18	block_groups	calculate
10	low income	<= $50,000		median hh income	block_groups	
11	serving the most people	<= 0.25 miles		population	block_centroids	
12	final map		political boundaries			
13	final map		roads			
14	final map		relief			

12) Choose the final map data. The last rows of the table list data that we may need for cartographic reasons in the final map.

Row 12 lists political boundaries. Our source data has feature classes of cities, counties, and states. Until we design our final map, we won't know for sure, but we foresee a likely need for city and county boundaries, and not for state boundaries.

Ⓐ In ArcMap, in the Catalog window, under ESRI.gdb, expand Boundary.

We have feature classes of city, county, and state boundaries. In Lesson 2, we previewed *City_ply* and saw that its extent covered the entire United States. That's more data than we'll need. We never previewed the *County* feature class, so let's do that now.

Ⓑ Under Boundary, right-click the *County* feature class and choose Item Description.

In the Item Description window, on the Description tab, the thumbnail graphic indicates that the data covers the United States. Thumbnail graphics aren't automatically synchronized with the data, though, and may offer an outdated snapshot. It's a good idea to preview the geography, which is always current.

Ⓒ In the Item Description window, click the Preview tab (Figure 3-32). Leave the Item Description window open.

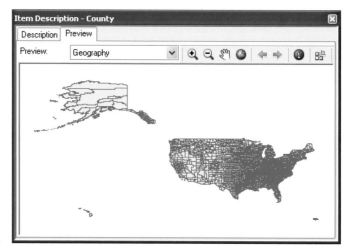

Figure 3-32

The preview confirms the spatial extent of the feature class.

Ⓓ In Row 12 of the requirements table, under DATASET, enter **City_ply**. Press Enter to start a new line and enter **County**.

Ⓔ In Row 12, under PREPARATION, enter **reduce extent** next to *City_ply* and *County*.

In Row 13, we have a need for roads listed.

F In ArcMap, in the Catalog window, under ESRI.gdb, expand Transport.

G Click the *Mjr_rd* feature class.

The Item Description window updates to preview the *Mjr_rd* geography.

H Click the Description tab and read the summary, then close the Item Description window.

The data covers the greater Los Angeles area, which is appropriate.

I In Row 13 of the requirements table, under DATASET, enter `Mjr_rd`.

Relief and imagery are available as online basemap layers.

J In Rows 14 and 15 of the requirements table, under DATASET, enter **basemap** (Figure 3-33).

#	REQUIREMENT	DEFINED AS	SPATIAL DATA	ATTRIBUTE DATA	DATASET	PREPARATION
1	land parcel		parcels		Parcels	
2	vacant			land use	Vacant Parcels	join table to Parcels
3	one or more acres			area	Parcels	calculate area
4	within city limits			city name	Vacant Parcels	
5	near LA river	<= 0.75 miles	rivers		LARiver	dissolve
6	away from parks	> 0.25 miles	parks		Parkland	load data and edit features
7	in a neighborhood		census unit		block_groups	
8	densely populated	>= 8,000 per sq mi		population density	block_groups	calculate
9	lots of kids	>= 25%		age under 18	block_groups	calculate
10	low income	<= $50,000		median hh income	block_groups	
11	serving the most people	<= 0.25 miles		population	block_centroids	
12	final map		political boundaries		City_ply County	reduce extent reduce extent
13	final map		roads		Mjr_rd	
14	final map		relief		basemap	
15	final map		imagery		basemap	

Figure 3-33

13) Save your work. All the datasets we need for the analysis have been specified, so we can save and close our work.

A Save and close the data requirements table.

B In ArcMap, zoom to the *Los Angeles* layer.

C Save the map document.

D Optionally, remove the Data Frame Tools toolbar.

▷ How? From the ArcMap main menu, choose Customize → Toolbars, then uncheck Data Frame Tools.

E If you are continuing to the next exercise now, leave ArcMap open. Otherwise, exit ArcMap.

In the next exercise, we'll take a step back from the project. Analysis operations take place within a particular coordinate system, and choosing that system is an important part of setting up the project. To make a good choice, you should have some background knowledge of what coordinate systems are and how they're managed in ArcMap.

Exercise 3b: Choose a coordinate system

A coordinate system is a mesh of perpendicular intersecting lines superimposed on a surface. The point of intersection of any two lines is a unique location, which can be specified with two values (a coordinate pair). The values are measurements from a given reference point in a given unit of measure. The unit of measure may be an angular unit, like degrees, or a length unit, like feet or meters.

Coordinate systems can be applied to any surface, but we're interested in the surface of the earth. If we represent the earth with a spherical model, like a globe, we have a curved surface to deal with. If we represent it with a flat model, like a map, the surface is flat.

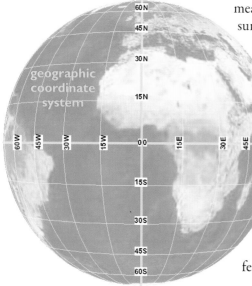

We apply a geographic coordinate system to round or roundish earth models. Geographic coordinate systems use an angular unit of measure because angles are better for measuring locations on a curved surface. We apply a projected coordinate system to flat earth models. Projected coordinate systems use a length unit of measure. Figure 3-34 illustrates both types.

Every usable spatial dataset has a coordinate system. The locations of its features are specified by coordinates that are correct within its own framework, but which would be wrong or absurd in another system. Fifty degrees of arc isn't the same as 50 meters, which isn't the same as 50 feet. Knowing the coordinates of a point doesn't tell you where the point is on the earth unless you also know what the coordinate system is.

There are many different geographic and projected coordinate systems. As long as two datasets use the same system, their features will align correctly on a map. Datasets with different

Figure 3-34

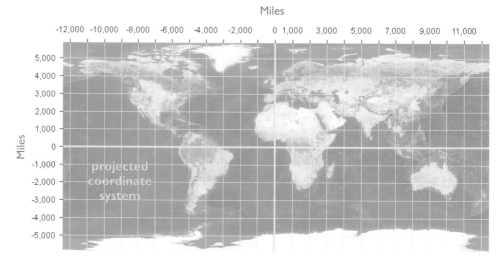

systems need to be reconciled. This basically means that one dataset's coordinate system has to be mathematically converted, or projected, to the other. This can be done with data processing tools. It is also done automatically by ArcMap when you add layers to a map document. What ArcMap reconciles, however, are the coordinate systems of the layers in the map, *not* the coordinate systems of the datasets that the layers point to. This reconciliation of coordinate systems in a map document is called "on-the-fly" projection.

We'll see how ArcMap manages on-the-fly projection of coordinate systems. We'll also look at how map display and map measurements change with different geographic and projected coordinate systems. Finally, we'll select a coordinate system for our analysis project.

1) **Check the coordinate system of a dataset**. Every spatial dataset has a coordinate system. Let's see how to find out what it is.

- Ⓐ If necessary, start ArcMap and open a blank map.
  - ▷ If ArcMap is already running, on the Standard toolbar, click the New Map File button ☐ and open a blank map.

- Ⓑ In the Catalog window, under ESRI.gdb, expand Boundary. Right-click the *State* feature class and choose Properties.

- Ⓒ In the Feature Class Properties dialog box, click the XY Coordinate System tab (Figure 3-35).

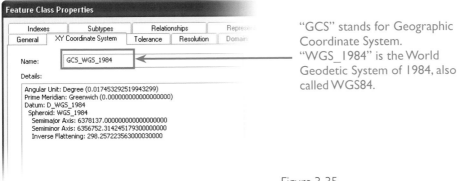

"GCS" stands for Geographic Coordinate System. "WGS_1984" is the World Geodetic System of 1984, also called WGS84.

Figure 3-35

The name of the coordinate system is GCS_WGS_1984. Under the name are details giving the essential properties of the system:

- Its unit of measure (degrees of angle)
- Its prime meridian (Greenwich)
- Its datum (D_WGS_1984)

The datum includes the name and dimensions of a particular spheroid, or roundish earth model.

WGS_1984 is a geographic coordinate system. Geographic coordinate systems, and datasets that use them, are said to be "unprojected." If we checked the other feature classes in ESRI.gdb, we'd find that they all have this same geographic coordinate system.

Ⓓ Close the Feature Class Properties dialog box.

2) Add data to ArcMap. Let's add this unprojected data to ArcMap and see what it looks like.

Ⓐ Add *State* as a layer to ArcMap (Figure 3-36).

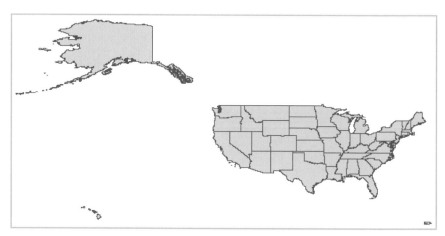

Figure 3-36

Ⓑ Open the layer properties for the *State* layer and click the Source tab.

The same coordinate system information (minus the details) that you saw in the feature class properties can be found in the layer properties as well (Figure 3-37).

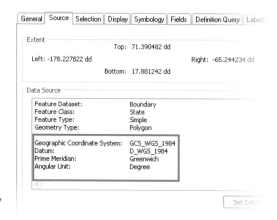

Figure 3-37

C Close the Layer Properties dialog box.

D Move the mouse pointer over the map (Figure 3-38).

In the lower-right corner of the application window, the coordinates of the mouse pointer are reported in decimal degrees. The first coordinate (longitude) tells you the angular position east or west of the prime meridian. The second coordinate (latitude) tells you the angular position north or south of the equator.

Figure 3-38

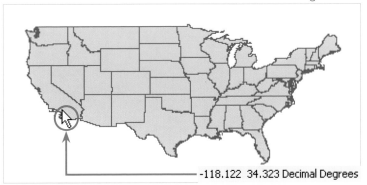

-118.122 34.323 Decimal Degrees

Negative values are used for west longitude (as here) and south latitude.

Degrees Minutes Seconds and Decimal Degrees

Angle measurements are commonly expressed in degrees, minutes, and seconds (DMS), where a circle has 360 degrees, a minute is 1/60th of a degree, and a second is 1/60th of a minute. ArcMap can express latitude-longitude values this way, but it makes calculations in decimal degrees (DD), which is a conversion of DMS to decimal notation. The following two expressions are equivalent:

-118.122, 34.323 (DD)

-118° 7' 19", 34° 19' 23" (DMS)

E From the ArcMap main menu, choose View → Data Frame Properties.

F In the Data Frame Properties dialog box, click the Coordinate System tab (Figure 3-39).

A data frame ≡ is a container for layers in a map document. Every map document starts with one data frame, called Layers, which is the top entry in the Table of Contents. You can insert more data frames to manage separate sets of data—essentially, multiple maps—within your map document. Why would you want to do that? It's usually a layout consideration: you may want to print a single map sheet that contains a main map, an overview map, an inset map, or some other combination of views. (We'll see how data frames work in map layouts in Lesson 8.)

One crucial function of a data frame is to enforce the spatial alignment of the layers it contains. An empty data frame, in a new map document, isn't associated with any coordinate system. When you first add a layer to it, it adopts that layer's coordinate system as a standard. Any layers added thereafter are converted (projected) by ArcMap into that same system. The process is called on-the-fly projection and happens automatically as new layers are added.

G Close the Data Frame Properties dialog box.

Figure 3-39

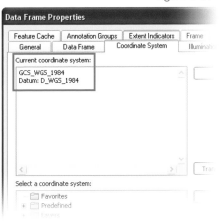

3) Project data on the fly. To see how on-the-fly projection works, we'll insert a second data frame, add a layer that has a different coordinate system from the *State* layer, then copy the *State* layer into the new data frame. To see the contents of both data frames at the same time, we have to open another viewer window.

Ⓐ On the Tools toolbar, click the Create Viewer Window tool 🔲 and drag a box around the data (Figure 3-40).

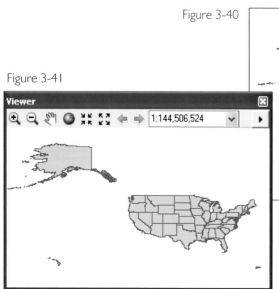

Figure 3-40

Figure 3-41

Move the viewer window by dragging its title bar. Resize it by dragging a side or corner.

When you release the mouse button, you'll have a viewer window (Figure 3-41) that you can move and resize however you like. Viewer windows are usually used to give you multiple vantage points on a map, but in this case, the purpose is to preserve a snapshot of the data as it looks in a geographic coordinate system.

At the top of the viewer window is a toolbar with navigation tools. Viewer windows can be zoomed and panned independently of the main map.

Ⓑ Experiment with panning and zooming in the viewer window. When you're finished, go back to the full extent ◉.

Ⓒ From the main menu, choose Insert → Data Frame.

A new data frame, called "New Data Frame," is added to the Table of Contents. Its name is boldfaced to indicate that it's the active data frame, the one we're working with. (The title bar of the viewer window tells you that the data frame called "Layers" is inactive.) The main map window is empty because the new data frame doesn't contain any layers.

Ⓓ From the main menu, choose View → Data Frame Properties.
   ▷ A shortcut is to double-click the data frame name in the Table of Contents.

Ⓔ If necessary, click the Coordinate System tab.

The current coordinate system is "No projection." We just wanted to confirm that a new data frame isn't associated with any coordinate system.

⑤ Close the Data Frame Properties dialog box.

⑥ Add the Imagery basemap (Figure 3-42).

▷ How? Click the Add Data drop-down arrow and choose Add Basemap. In the dialog box, click Imagery.

⑦ Move the mouse pointer over the map.

The coordinates, in the lower-right corner of the ArcMap window, are in meters. That's a sign that you're working with a projected coordinate system, one based on length measurements rather than angles.

⑧ Open the data frame properties again.

Figure 3-42

The data frame's coordinate system, adopted from the *Basemap* layer, is now WGS_1984_Web_Mercator_Auxiliary_Sphere (Figure 3-43). A projected coordinate system essentially consists of a map projection (in this case, a form of the Mercator projection) applied to a geographic coordinate system (in this case, GCS_WGS_1984). See the topics *Geographic coordinate systems* and *Projected coordinate systems* on pages 110–111 for more information.

⑨ Close the Data Frame Properties dialog box.

⑩ In the Table of Contents, under the inactive Layers data frame, right-click the *State* layer and choose Copy.

⑪ Right-click on the name New Data Frame and choose Paste Layer(s).

⑫ Under New Data Frame, right-click the *State* layer and choose Zoom To Layer.

The *State* layer aligns with the *Basemap* layer (Figure 3-44). In the process, its spatial properties have changed. Compare its appearance in the main map window and the viewer. Notice especially the shape of Alaska.

⑬ Close the viewer window.

Current coordinate system:

WGS_1984_Web_Mercator_Auxiliary_Sphere
Projection: Mercator_Auxiliary_Sphere
False_Easting: 0.000000
False_Northing: 0.000000
Central_Meridian: 0.000000
Standard_Parallel_1: 0.000000
Auxiliary_Sphere_Type: 0.000000
Linear Unit: Meter

GCS_WGS_1984
Datum: D_WGS_1984

Figure 3-43

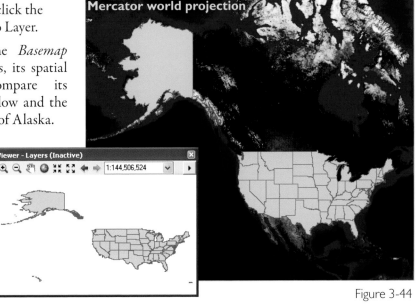

Figure 3-44

4) Make an area measurement. Coordinate systems change measurements as well as appearances. The Mercator projection we're using now is a conformal projection, which means that it shows shapes correctly. (Look at Alaska on a globe and you'll see this is true.) On the other hand, it distorts area measurements—pretty significantly in extreme latitudes.

Ⓐ Zoom in on Alaska. (It doesn't matter how far exactly.)

Ⓑ On the Tools toolbar, click the Measure tool 📏.

Ⓒ At the top of the Measure dialog box, click the Clear and Reset Results button ×, then click the Measure A Feature button +.

Ⓓ Click the Choose Units button ▾, point to Area, and choose Miles.

Ⓔ Click anywhere inside Alaska (Figure 3-45). Leave the Measure dialog box open.

Figure 3-45

The area measurement for Alaska is 3,142,683 square miles.

Ⓕ Under New Data Frame, open the *State* layer attribute table.

Ⓖ Select the record for Alaska (Figure 3-46). (It should be the second one.) Scroll all the way to the right to see the SQMI attribute.

Click the gray square to select the record.

Figure 3-46

Table

NO_FARMS97	AVG_SIZE97	CROP_ACR97	AVG_SALE97	SQMI	Shape_Length	Shap
41384	210	4197670	74.86	51658	20.841598	1
548	1608	94810	44.96	581385	553.015469	279
6135	4379	1277169	310.25	113998	23.915246	2
45142	318	10062289	121.39	53179	22.048003	1
74126	374	10803804	310.72	158149	55.601531	4
28268	1154	10509384	160.4	104094	22.021594	
3687	97	181043	114.36	4975	10.668557	

The value in the table, which is correct, is 581,385 square miles. (You can find other values cited in reference sources, but they're similar to this one.) That means that the map measurement is an exaggeration by a factor of more than five.

Figure 3-47

The data frame's current coordinate system

H At the top of the table window, click the Clear Selection button 🔲 . Close the table.

The Measure tool is doing its job correctly: it has measured the area of Alaska as distorted by the Mercator world map projection. (The Mercator projection is responsible for some entrenched misconceptions about the relative sizes of land masses.)

5) Change the data frame's coordinate system. You can change the data frame's coordinate system whenever you want—before or after adding layers.

A Open the data frame properties for New Data Frame.

On the Coordinate System tab, the bottom window contains four folders as shown in Figure 3-47:

- *Favorites* is for coordinate systems that you use often. (It's empty until you add to it.)
- *Predefined* contains the 4,500 or so coordinate systems ArcMap supports.
- *Layers* shows the "native" coordinate system (the one belonging to the layer's source data) of each layer in the data frame.
- *<custom>* shows the current coordinate system and any others that you may have modified (for example, by changing reference lines or measurement units).

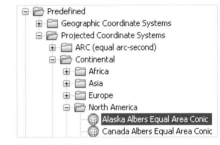

Figure 3-48

B Expand the Predefined folder, then the Projected Coordinate Systems folder, then Continental, then North America.

The first coordinate system listed in this folder should be Alaska Albers Equal Area Conic.

C Click on Alaska Albers Equal Area Conic to highlight it (Figure 3-48), then click OK.
  ▷ If you get a warning, click Yes to use the coordinate system anyway.

The new coordinate system is applied to the data frame. Both the *Basemap* layer and the *State* layer are projected (or reprojected, if you prefer) on the fly. Your map should look like Figure 3-49.

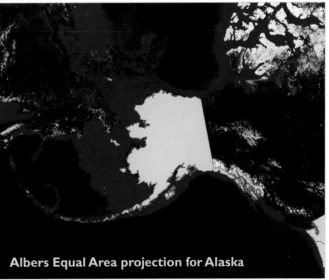

Albers Equal Area projection for Alaska

Figure 3-49

Geographic coordinate systems

A geographic coordinate system (GCS) is based on a spheroidal model of the earth. Sometimes a perfect sphere is used, but usually a slightly squashed "oblate spheroid" is used to reflect the fact that the earth bulges at the equator and is flattened at the poles.

Sphere   Oblate spheroid
                (much exaggerated)

The reference lines of a geographic coordinate system are parallels and meridians. Parallels are lines that circle the globe parallel to the equator. Meridians are lines perpendicular to the equator that converge at the poles. By convention, the origin of the system (its 0,0 coordinate), is the intersection of the equator and the prime meridian, the meridian passing through Greenwich, England.

Geographic coordinates, commonly called latitude-longitude values, are measurements of angle, not distance. Angles are a constant unit of measure on a sphere, while distances are not (because meridians converge).

Angle measurements are usually expressed in degrees, minutes, and seconds. A degree has 60 minutes; a minute has 60 seconds.

Latitude is angular position north or south of the equator. The equator is 0° latitude, the North Pole is 90° north and the South Pole is 90° south.

Longitude is angular position east or west of the prime meridian. The prime meridian is 0° longitude. Its anti-meridian (on the other side of the world) is both 180° east and 180° west.

A latitude-longitude pair defines a unique position on the earth's surface.

The unique location of Dodger Stadium, for example, would be written like this: **34°4'26"N, 118°14'27"W** and spoken like this:

"34 degrees, 4 minutes, 26 seconds north latitude; 118 degrees, 14 minutes, 27 seconds west longitude."

For computer calculations, these values are converted to decimals. The location of Dodger Stadium in "decimal degrees" is 34.073, -118.24. The minus sign is used for west longitude and south latitude.

The fact that there are many different geographic coordinate systems is a source of trouble for GIS users. What makes two systems different is disagreement about the exact latitude-longitude values of particular locations. Why is there disagreement about that? In simple terms, it's because different spheroid models of the earth have been developed over time by different earth scientists using different technologies. Changing the shape or size of the model ends up changing the coordinates of points on the surface—usually not very much, but sometimes, in sensitive applications, enough to be of concern. This issue is taken up in more detail in the *Datums* topic later in this lesson.

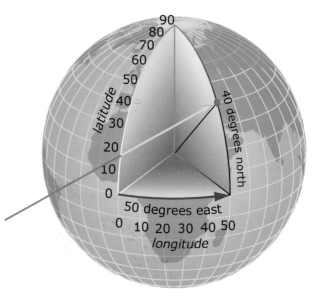

Projected coordinate systems

The earth's surface can be modeled very well on a spheroid, but not as well (except over small areas) on a plane. To make a map, you more or less have to flatten a sphere, which is like squaring a circle, only harder. It can't be done without radically adjusting the spatial properties and relationships of features on the surface: their shapes, sizes, and relative distances and directions.

The name for any such radical adjustment is a map projection. A projection is a mathematical formula (there are lots of different ones) for translating the world into flat space. All map projections introduce spatial distortion. They are variously designed to minimize certain kinds of distortion or to distribute it in certain ways over the map surface. Some projections correctly preserve feature shapes but distort their areas. Some preserve areas but distort shapes. Some compromise. Some have special properties, like keeping true distance measurements from a single point to all others, or ensuring that courses of constant compass bearing are plotted as straight lines. The smaller the area being mapped, the less distortion there is of any spatial properties. Areas up to medium-sized countries (say about the size of Nigeria or Bolivia) can be mapped with distortion so low as to be insignificant.

A projected coordinate system consists of a map projection, a length-based unit of measure, an origin point for measurements, and other parameters, such as standard lines that define the distortion pattern on the map. Manipulating these parameters is what allows you to customize a coordinate system for a specific area of interest. Because a projection is applied to a particular spheroid and its definition of latitude-longitude values, a projected coordinate system also includes, or presupposes, the geographic coordinate system from which it is derived.

The idea of projection includes both going from a geographic (unprojected) system to a projected system, and going from one projected system to another (sometimes called reprojection). To go from one projected system to another, ArcMap undoes the map projection, goes back to the underlying geographic coordinate system, and applies a new projection to it. ArcMap stores thousands of map projection formulas and can run these calculations quickly.

Some common world map projections (and a couple of uncommon ones)

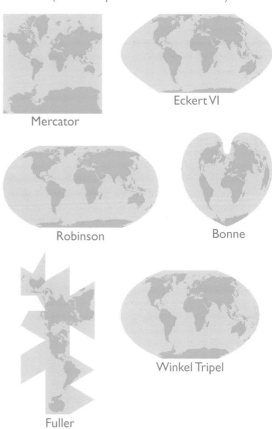

Mercator

Eckert VI

Robinson

Bonne

Fuller

Winkel Tripel

When you add an unprojected dataset to a map (that is, a dataset that stores feature coordinates as latitude-longitude values), the data still has to be projected in some sense in order to be viewed as a map. In ArcMap, this default "pseudoprojection" has the display properties of a map projection (specifically, the Plate Carree), but none of the other properties or parameters of a projected coordinate system.

geographic pseudoprojection

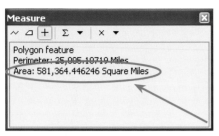

Figure 3-50

6) **Measure Alaska again.** Equal-area projections (of which the Albers Equal Area Conic is one) preserve the spatial property of area, or size. The trade-off is that they don't represent shapes correctly. Because this projection is customized for Alaska, however, most distortion of *all* types is pushed outside the area of interest.

Ⓐ Click on the Measure dialog box to make it active.

▷ Make sure the Measure A Feature tool is still selected.

Ⓑ Click on Alaska.

The area is given accurately as 581,364 square miles (Figure 3-50).

Ⓒ Close the Measure dialog box.

Ⓓ Zoom to the full extent (Figure 3-51).

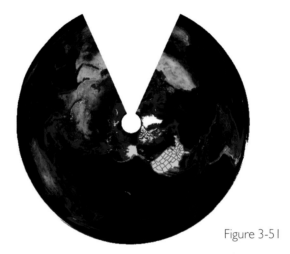

Figure 3-51

This coordinate system is obviously unsuitable for a world map. It doesn't show the whole world for one thing. The price for mapping Alaska with very low distortion is that places like Australia and South America are severely distorted—which is fine, as long as we keep our map zoomed in to Alaska.

Let's summarize how things stand with the layers in this data frame:

• The geographic coordinate system of the *State* layer is GCS_WGS_1984. It has no projected coordinate system.

• The projected system of the *Basemap* layer is Web_Mercator. Its underlying geographic system is GCS_WGS_1984.

• The data frame has been set to the Alaska Albers Equal Area Conic projected system. Its underlying geographic system is GCS_North_American_1983.

• The *State* and *Basemap* layers have been projected on the fly into the Alaska Albers system.

No matter how often we change coordinate systems, ArcMap keeps the data in alignment.

7) **Add another layer.** Let's get back to our true area of interest.

Ⓐ In the Catalog window, from the City of LA folder, drag and drop *LARiver* onto the map.

Ⓑ Zoom to the *LARiver* layer and turn off the *State* layer.

Ⓒ Change the color of the *LARiver* symbol to Cretean Blue (Figure 3-52). Change its line width to 2.

A coordinate system meant for Alaska (or the world) isn't appropriate for Southern California. Local mapping needs across the United States are served by a system called the State Plane Coordinate System. This isn't a single coordinate system for the entire country, but a patchwork of systems, each of which covers a state or part of a state. Together, they assure that for whichever part of the country you want to map, you get uniformly low distortion. California is divided into six State Plane zones, with Los Angeles falling in Zone 5 (Figure 3-53).

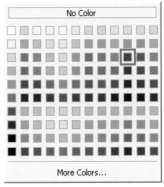

Figure 3-52

California Zone 5

Figure 3-53

The State Plane Coordinate System divides states into zones and applies a locally optimized coordinate system to each zone. Vertical zones are based on the Transverse Mercator projection. Horizontal zones are based on the Lambert Conformal Conic projection.

8) **Change the data frame's coordinate system.** We'll change the data frame's coordinate system again, this time to State Plane California Zone 5.

Ⓐ Open the data frame properties for New Data Frame.

Ⓑ On the Coordinate System tab, in the Select a coordinate system box, expand the Layers folder, and expand the LARiver folder.

This is still one more way—along with feature class properties and layer properties—to find the native coordinate system of a dataset.

Ⓒ Click the coordinate system name inside this folder (Figure 3-54).

Figure 3-54

This resets the current coordinate system of the data frame (Figure 3-55). The State Plane California Zone 5 system is a projected coordinate system based on a Lambert Conformal Conic map projection. Its underlying geographic coordinate system is North American 1983. This is different from the WGS 1984 system used by both the *State* and *Basemap* layers.

ⓓ Click OK on the Data Frame Properties dialog box.

A warning appears (Figure 3-56). It's a variation of the geographic coordinate systems warning we've seen before. Both warnings alert you to a mismatch between the geographic coordinate system of the data frame and one or more of its layers. (You get the warning shown here when you change the data frame's coordinate system; you get the other warning when you add a nonmatching layer.)

Figure 3-55

Figure 3-56

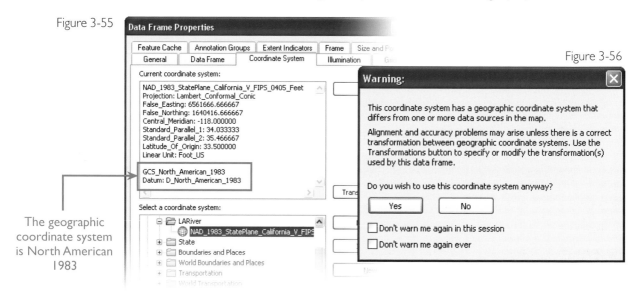

The geographic coordinate system is North American 1983

Why does ArcMap keep issuing these warnings, and how serious are they? Perfect projection or reprojection of data into a different coordinate system requires that the "from" and "to" systems have the same geographic coordinate system. If they don't, it means they have a basic disagreement about the latitude-longitude values of locations. Projection can't fix that disagreement. Projection converts a given set of angular coordinates to planar coordinates. It doesn't convert one set of angular coordinates to another.

In this exercise, and elsewhere in the book, we've been mixing layers based on two geographic coordinate systems: WGS 1984 and North American 1983. Whenever we do so, ArcMap lets us know that it can't guarantee perfect alignment of the data. The disagreements between these two systems are very small (they rarely translate into more than a few feet on the ground), but ArcMap doesn't know what we're doing with the data or how much accuracy we need. There is a way to reconcile different geographic coordinate systems—it's called a geographic transformation or a datum transformation—but we'll postpone doing one until the next lesson.

E Click Yes on the warning.

Your map should look like (Figure 3-57). The State Plane California Zone 5 coordinate system is the one we'll use for our analysis.

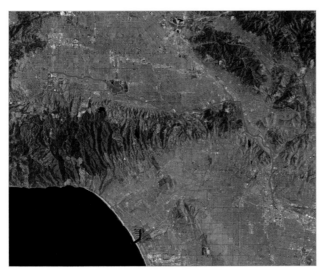

Figure 3-57

F Save the map document as **Lesson3b** in the MapsAndMore folder.

G Exit ArcMap.

Geographic coordinate systems warnings

When you get a geographic coordinate systems warning, check boxes allow you to turn them off, either for the duration of the ArcMap session or for all future sessions. Unless you're an expert, you shouldn't check "Don't warn me again ever." If you do so by mistake, you can reinstate the warning by opening AdvancedArcMapSettings.exe in the ArcGIS\Desktop10.0\ Utilities folder. In the Advanced ArcMap Settings Utility, click the TOC/Data tab and uncheck the box "Skip datum check." You need power user or administrative rights to make this change, which takes effect in the next ArcMap session.

In the next lesson, we'll resume work on the project by taking care of the preparation tasks we listed in the data requirements table.

Datums

A geographic coordinate system is defined by three things: an angular unit of measure (usually degrees), a prime meridian (usually Greenwich), and a datum. The datum is the part that gives people trouble. To understand it, let's start with the shape of the earth.

The earth isn't a perfect sphere, or even a mathematically regular spheroid. It's a lump with an uneven shape owing to different concentrations of mass (and therefore unequal gravity) over its surface. In addition, it has topographic features like mountains and valleys.

When the spatial positions of features are determined—as was formerly done by survey, and is now mostly done by satellite—they are first determined on the earth's surface. These raw measurement values are then mathematically "leveled" to a geoid. A geoid is the (still gravitationally lumpy) shape that the earth would have if it were covered by a mean sea level surface—in other words, if it had no topography.

The shape of the geoid, however, is too complex to be a working model. So the next step is to move the measurements from the geoid to a spheroid: a model with a regular, non-lumpy shape.

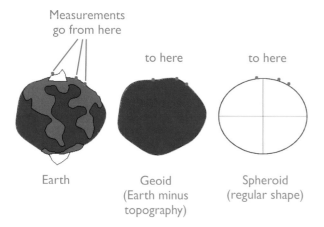

Measurements go from here

to here

to here

Earth

Geoid (Earth minus topography)

Spheroid (regular shape)

That's where the datum comes in. The datum is two things: first, it's a chosen spheroid, which could be WGS 1984, GRS 1980, Clarke 1866, Bessel 1841, or a number of others. (The world is standardizing on the GRS 1980 spheroid, but isn't all the way there yet.) Second, it's a mathematical orientation, or "fit," of the geoid to the spheroid. In the transfer of measurements from geoid

to spheroid, some error will be introduced because the lumps have to be smoothed out. How that error is distributed is the "fit." One approach is to make the fit really good for one part of the world, like North America, and not to worry about the rest. That's a local datum. It's designed to maintain high accuracy for measurements over a limited area. The other approach is to average the error over the whole surface. That's an earth-centered datum. It's designed to maintain high accuracy for the world as a whole.

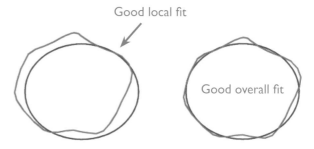

Good local fit

Good overall fit

When two geographic coordinate systems are different, it's usually because the datums are different (which, in turn, is either because the spheroids are different or the fit is different). When you get a coordinate systems warning in ArcMap, one possibility is to ignore it and leave the data slightly out of alignment. Depending on your needs for accuracy, this may be an entirely sensible choice. The amount of misalignment depends on the datums involved and the part of the world being mapped, but in the mismatch North Americans usually encounter (between the World Geodetic System of 1984 and the North American Datum of 1983), it typically doesn't exceed a few feet. At most scales, this isn't noticeable.

The other option is to reconcile the systems through a geographic transformation. Transformations are often done in conjunction with a coordinate system projection. Like projections, they can be permanently applied to datasets with data processing tools, or they can be done on the fly in ArcMap. They require some expert knowledge. There are default methods to convert one spheroid to another, but there aren't default fits, because the right fit depends on your area of interest. The document *geographic_transformations.pdf*, in the ArcGIS\Desktop10.0\Documentation folder, can help you find the right fit for an area of interest.

Lesson

4 Build the database

WE'VE COME TO A TURNING POINT

in our project. In the early, exploratory phases, we looked at data—both in and out of maps—and evaluated it from various standpoints. We worked a lot with layer properties, but we didn't operate directly on the data. We'll do that now.

We have three main tasks in this lesson. First, we'll create a geodatabase to store the data for the project. That raises the question whether the project really needs its own database. Couldn't we continue to use the collection of folders and files we've been working with so far? Yes, we could, but there are good reasons to make a new database:

- We can keep the project data in a single location, rather than having it distributed among various workspaces.
- We can keep the project data separate from datasets that we don't need.
- We can streamline the project data by eliminating unnecessary features and attributes, while maintaining the original source data as a backup.
- We can impose uniform standards like a common data format and consistent names for datasets.
- It will be easier to share the project data with others and to reuse it ourselves.

Our second task will be to populate the geodatabase with the data that we chose in the last lesson. This will be our introduction to the geographic data processing or "geoprocessing" tools that make ArcGIS Desktop so powerful.

Our third task will be to complete the preparation tasks in the data requirements table. Besides using geoprocessing tools, we'll do spatial queries, table calculations, and other operations.

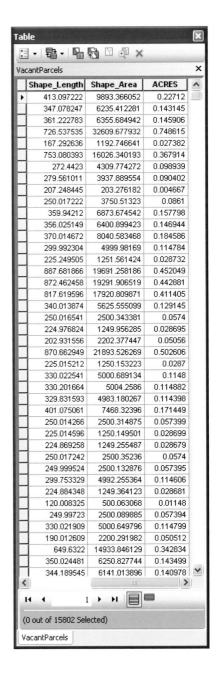

Lesson Four roadmap

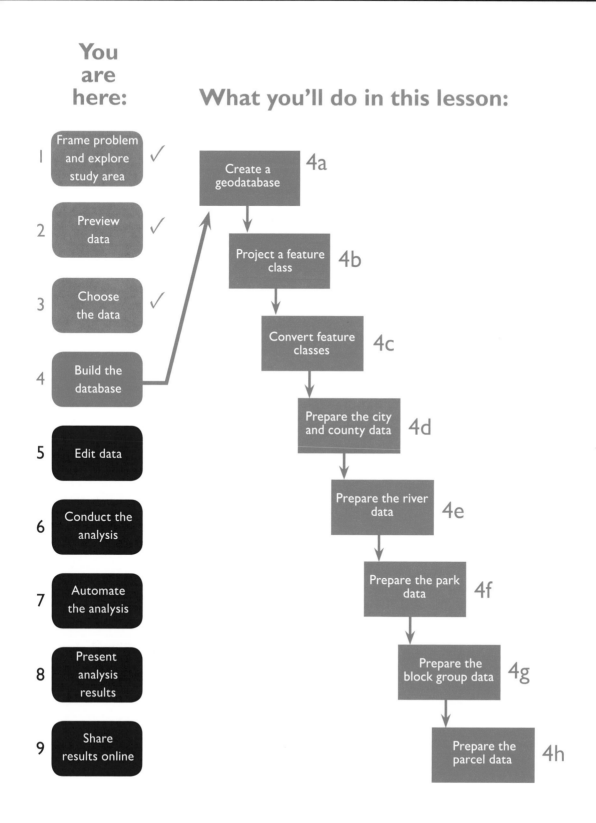

You are here:

What you'll do in this lesson:

1   Frame problem and explore study area ✓

2   Preview data ✓

3   Choose the data ✓

4   Build the database

5   Edit data

6   Conduct the analysis

7   Automate the analysis

8   Present analysis results

9   Share results online

Create a geodatabase   4a

Project a feature class   4b

Convert feature classes   4c

Prepare the city and county data   4d

Prepare the river data   4e

Prepare the park data   4f

Prepare the block group data   4g

Prepare the parcel data   4h

Exercise 4a: Create a geodatabase

In this short exercise, we'll create the database for our project. Our database is a simple one, but it still requires some thought about data format, organization of inputs and outputs, naming conventions, and other things. Our considerations are laid out in the topic *Project database considerations* on page 122.

1)   Create a geodatabase. We'll work in ArcMap in this lesson, but we'll use the Catalog window a lot for data management tasks.

A Start ArcMap and open a blank map.

Inside the ParkSite folder, we'll create a new folder for the project. Eventually, it will contain a few geodatabases and some other things, but in this lesson, it will just have one geodatabase with the starting data for the project.

B In the Catalog window, expand Folder Connections, then expand the C:\UGIS folder.

C Right-click the ParkSite folder. On the context menu, point to New and choose Folder.

D Name the folder **AnalysisData** and press Enter.

E Right-click the AnalysisData folder, point to New, and choose File Geodatabase.

F Rename the geodatabase **ReadyData** and press Enter.

Every map document has a default geodatabase, to which ArcMap directs geoprocessing output data (see page 93). Unless otherwise specified, the default geodatabase is located in your \MyDocuments\ArcGIS folder. (For Windows 7 users, the folder is \Documents\ArcGIS.)

G Under AnalysisData, right-click the ReadyData geodatabase and choose Make Default Geodatabase.

In the Catalog window, the name of the geodatabase is boldfaced and a "home" is added to the icon (Figure 4-1). Right now, ReadyData is just an empty container. In the next exercise, we'll start to fill it with data.

Figure 4-1

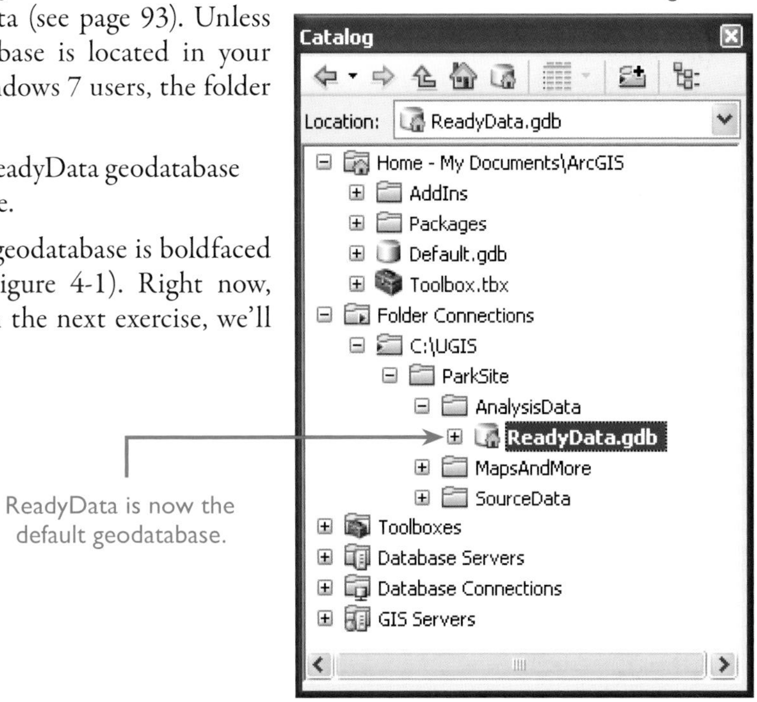

ReadyData is now the default geodatabase.

Project database considerations

For our simple database, we considered the following issues:

Data format: We put all the data into file geodatabase format. The geodatabase is the standard Esri spatial data format, and has many advantages, large and small, over the shapefile format. We decided not to use feature datasets 🖳. Feature datasets are only required to support advanced data structures (topologies and networks) that aren't discussed in this book. They can also be used as thematic organizers (they're used this way in the ESRI geodatabase in the source data for this project), but we won't have enough feature classes in our project database to make that worthwhile.

Inputs and outputs: An analysis project has input and output data. We'll store our input, or starting, data in one geodatabase and our output data in a separate geodatabase. This isn't the only way to do it. We could store both the input and output data in the same geodatabase. The reason we're doing it this way is because eventually we might want to run variations on the analysis. This would lead to multiple sets of results, where each set could consist of several feature classes. Keeping track of all that data in one geodatabase would be confusing.

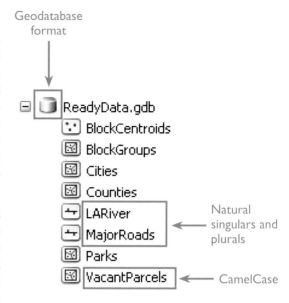

Geodatabase format

ReadyData.gdb
BlockCentroids
BlockGroups
Cities
Counties
LARiver — Natural singulars and plurals
MajorRoads
Parks
VacantParcels — CamelCase

Naming conventions: Conventions aren't right or wrong in and of themselves. Ideally, they should be easy to remember and easy to apply, but the only real sign that a convention is "wrong" is that you don't stick to it. Our conventions are:

- To separate word forms with capital letters, rather than spaces or underscores. For example, *VacantParcels* rather than *Vacant_Parcels* or *Vacant Parcels*. (This typographic convention is called "Pascal case" or "Camel case.")

- To use natural singulars and plurals. For example, *LARiver* (one river) and *MajorRoads* (many roads).

- To use complete names, not abbreviations. For example, *MajorRoads*, not *MjrRds*.

- To prefer descriptive names to names that reflect software procedures (the ArcMap default). For example, *ExcludedParkAreas* is better than *Parks_Buffer_Dissolve_Erase*.

Coordinate system: Feature classes in a geodatabase don't have to be in the same coordinate system. (Feature classes in the same feature dataset do have to be in the same system.) Since ArcMap projects data on the fly, you don't have to worry too much about mixing and matching datasets. One reason, however, to keep all the data in the same coordinate system is that we don't have to think about making sure the data frame is set to the appropriate system. Also, it's considered better practice to make spatial edits to data in its native coordinate system than when it's projected on the fly. We'll project all the data in our database to the State Plane California Zone 5 coordinate system.

Maintenance and documentation: We don't have to worry about updates, security, quality control, or other issues that apply to keeping a large database up and running. For us, maintaining the database means keeping it uncluttered. In the course of an analysis project, it's common to create datasets that are an important link in a chain of processes, but that have no particular value once the end of the chain is reached. If you don't remove this "intermediate" data from the geodatabase, it accumulates quickly, often making it hard to distinguish critical outputs and results from byproducts. (This is especially true for others who may want to look at or use your data, but it's true for you, as well.) In addition to keeping the database trim, we also want to keep it reasonably well documented. Output datasets that we intend to keep should have their item descriptions updated to reflect the contents and purpose of the data.

Types of geodatabases

A geodatabase can be one of three types:

A **file geodatabase,** with extension .gdb, is stored as a collection of system files in a folder. It has no total size limit. Each dataset within the geodatabase has a default size limit of one terabyte, which can be increased to 256 terabytes. Multiple users can concurrently edit different datasets within the geodatabase.

A **personal geodatabase,** with extension .mdb, is a Microsoft® Access® database. It has a total size limit of two gigabytes. Only one user at a time can edit the database. File and personal geodatabases serve the same purposes; the personal geodatabase is just an older version of the technology.

An **ArcSDE geodatabase** is stored in a relational database system such as Oracle®, Microsoft SQL Server™, or IBM® DB2® and is managed through software called ArcSDE® (SDE stands for Spatial Database Engine). Its size limits are determined by the database system in which it resides. ArcSDE geodatabases support concurrent editing on individual datasets and provide database administration tools.

Exercise 4b: Project a feature class

Our second goal is to populate the geodatabase with starting data for the analysis. As needed, we'll convert the data format from shapefile to geodatabase and project it to the State Plane coordinate system.

1)   Open the data requirements table. Having this table open may help you keep the various preparation tasks in mind. As you work through the lesson, you can keep it open or not as you prefer.

- Ⓐ Using Windows Explorer, open the data requirements table from the MapsAndMore folder.
  - ▷ If your table isn't up-to-date, download the results for Lesson 3 from the Understanding GIS Resource Center.

Most of the datasets need some preparation. Actually, they all do, if we count format conversion and projection as preparation tasks. Let's start with a dataset that doesn't have anything listed in the PREPARATION column. The *Mjr_rd* feature class in Row 13 is one.

- Ⓑ Minimize the data requirements table.

2)   Open the Project tool. First of all, let's check the coordinate system of the *Mjr_rd* feature class to make sure it needs to be projected.

- Ⓐ In ArcMap, in the Catalog window, expand the SourceData folder, the ESRI geodatabase, and the Transport feature dataset.
- Ⓑ Right-click *Mjr_rd* and choose Properties.

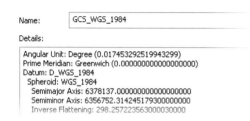

Name: GCS_WGS_1984

Details:

Angular Unit: Degree (0.0174532925199943299)
Prime Meridian: Greenwich (0.000000000000000000)
Datum: D_WGS_1984
 Spheroid: WGS_1984
  Semimajor Axis: 6378137.000000000000000000
  Semiminor Axis: 6356752.314245179300000000
  Inverse Flattening: 298.257223563000030000

Figure 4-2

C In the Feature Class Properties dialog box, click the XY Coordinate System tab (Figure 4-2).

The coordinate system is GCS_WGS_1984. We'll project it to the State Plane system. This isn't an on-the-fly projection of a layer in a data frame, but rather a geoprocessing operation on the feature class itself that converts all the WGS_1984 angular coordinates to State Plane linear coordinates. We need a geoprocessing tool to do it.

D Close the Feature Class Properties dialog box.

E In the Catalog Window, at the same level as Folder Connections, but underneath it, expand the Toolboxes folder. In the Toolboxes folder, expand System Toolboxes.

The system toolboxes contain the geoprocessing tools that come with ArcGIS. There are a lot of them, but most are pretty specialized. The ones most useful to our project (and to much typical GIS work) are the analysis, conversion, and data management tools (Figure 4-3).

F Expand Data Management Tools. Expand the Projections and Transformations toolset, then the Feature toolset.

As shown in Figure 4-4, toolboxes contain toolsets, which contain tools (and sometimes other toolsets).

Figure 4-3

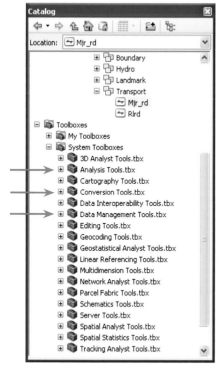

Figure 4-4

ⓖ Right-click the Project tool (marked with a hammer icon) and choose Open.

ⓗ In the lower right corner of the Project tool dialog box, click Show Help (Figure 4-5).

If you click in a parameter, the Help panel will update with contextual help.

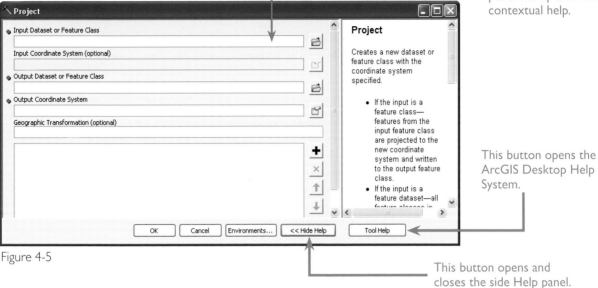

This button opens the ArcGIS Desktop Help System.

This button opens and closes the side Help panel.

Figure 4-5

Most tools look basically the same as this one: a dialog box with several settings, called parameters. Required parameters, like the input and output datasets, are marked with a green dot.

The Help panel summarizes the tool's function. This tool takes an input dataset and creates a new output dataset in whatever coordinate system you choose.

ⓘ Optionally, click Hide Help to close the Help panel.

3) Project the *Mjr_rd* feature class. Now we'll fill out the parameters and run the tool. The first thing to specify is the input dataset, the *Mjr_rd* feature class.

ⓐ In the Catalog window, scroll up to ESRI.gdb in the SourceData folder. Drag and drop *Mjr_rd* from the Catalog window to the Input Dataset or Feature Class parameter of the Project tool (Figure 4-6).

Figure 4-6

Once you drop the input dataset in place, ArcMap can fill out some of the other parameters for you. It reads the input coordinate system (GCS_WGS_1984) from the metadata and directs the output to the default geodatabase (Figure 4-7). This is where we want the data to go, but it's not the feature class name we want.

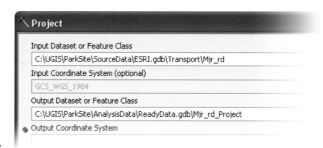

Figure 4-7

B In the Output Dataset or Feature Class box, highlight the entire path and press the Delete key. In its place, type **MajorRoads**.

C Press the Tab key.

The output path to the ReadyData geodatabase is restored.

D Next to the Output Coordinate System parameter, click the Spatial Reference Properties button.

E In the Spatial Reference Properties dialog box, make sure the XY Coordinate System tab is selected.

F Click Select. In the Browse for Coordinate System window, navigate to Projected Coordinate Systems → State Plane → NAD 1983 (US Feet) (Figure 4-8).

G Open this folder and click on NAD 1983 StatePlane California V FIPS 0405 (US Feet).prj. Click Add.

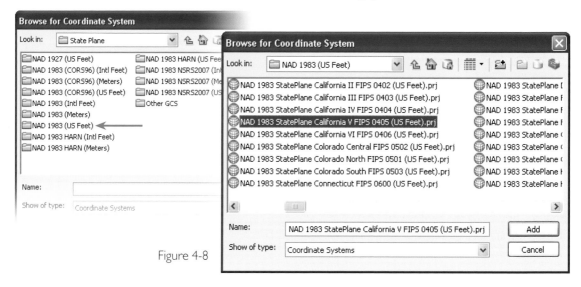

Figure 4-8

The Spatial Reference Properties dialog box shows the coordinate system details. The map projection is Lambert Conformal Conic. The values for the central meridian and standard parallels focus the coordinate system on a particular swath of southern California, pushing virtually all distortion outside the area of interest. Notice that the underlying geographic coordinate system is GCS_North_American_1983 (Figure 4-9).

Figure 4-9

Projection: Lambert_Conformal_Conic
False_Easting: 6561666.666667
False_Northing: 1640416.666667
Central_Meridian: -118.000000
Standard_Parallel_1: 34.033333
Standard_Parallel_2: 35.466667
Latitude_Of_Origin: 33.500000
Linear Unit: Foot_US (0.304801)

Geographic Coordinate System: GCS_North_American_1983
Angular Unit: Degree (0.017453292519943295)
Prime Meridian: Greenwich (0.000000000000000000)
Datum: D_North_American_1983

ℍ Click OK on the Spatial Reference Properties dialog box.

Now there's a green dot by the Geographic Transformation parameter. It's no longer optional because the input and output datasets have different geographic coordinate systems: WGS_1984 and NAD_1983, respectively. When you project layers on the fly, you can bypass geographic transformations, but the Project tool makes you do them. For more information about geographic transformations, see the topic *Datums* on page 116.

North American 1983 is the ArcGIS name for the geographic coordinate system based on the North American Datum of 1983 (NAD83).

ℐ Click the Geographic Transformation drop-down arrow and select NAD_1983_To_WGS_1984_5.

This looks backwards because we want to transform a WGS_1984 input to a NAD_1983 output, but the transformation works in either direction. (This is noted in the tool help panel if you have it open.) ArcMap sets the "from-to" relationship correctly by looking at the input and output datasets. Among the available variations, or "fits," NAD_1983_To_WGS_1984_5 is best for the contiguous United States.

ℐ Check your Project tool settings against Figure 4-10. Click OK.

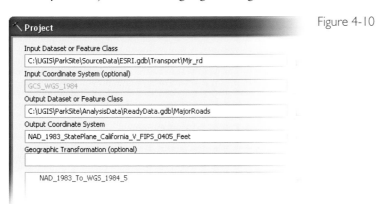

Figure 4-10

After a moment, a message scrolls along the bottom of the ArcMap window to show that the tool is running. When the process is done, a notification appears and fades in the lower right corner of your monitor.

The input feature class has been projected and saved as a new feature class to the ReadyData geodatabase with the name you gave it. A corresponding layer is added to the map (Figure 4-11). The original *Mjr_rd* feature class remains unchanged in the ESRI geodatabase.

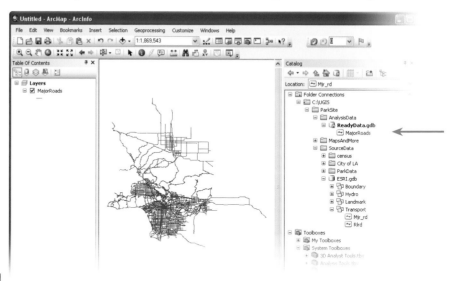

Figure 4-11

🅚 Confirm that the coordinate system of *MajorRoads* is State Plane California Zone V, and that its geographic coordinate system is GCS_North_American_1983.

▷ How? Use the feature class properties (XY Coordinate System tab) or the layer properties (Source tab).

🅛 From the ArcMap main menu, choose Geoprocessing ➜ Results.

🅜 In the Results window, expand Current Session and Project. Under Project, expand Inputs and Messages (Figure 4-12).

The Results window keeps a log of which geoprocessing operations have been run, at what times, and with what parameters. Results are especially useful when a process fails because the error is reported in the Messages section.

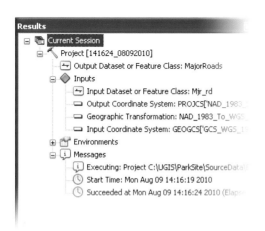

Figure 4-12

Double-clicking the tool icon (in this case, the hammer) next to the tool name will reopen the tool with the parameters you used. This can be a convenient way to rerun the tool if you just want to change a parameter.

By default, geoprocessing results are stored for two weeks. You can change this value by choosing Geoprocessing ➜ Geoprocessing Options from the main menu.

🅝 Close the Results window.

In the next exercise, you'll add two more feature classes to the ReadyData geodatabase.

Exercise 4c: Convert feature classes

In this exercise, we'll project the *block_centroids* and *Parkland* feature classes. We'll learn a couple of useful things. The first is how to set the geoprocessing environment. Environment settings are default values or conditions applied automatically to all geoprocessing operations. The second thing is how to use a batch tool that can process multiple inputs.

1) Set the geoprocessing environment. There are many environment settings, but we'll concentrate on the output coordinate system.

Ⓐ From the main menu, choose Geoprocessing → Environments.

Ⓑ In the Environment Settings box, click Output Coordinates to expand the heading (Figure 4-13).

Ⓒ For the Output Coordinate System parameter, click the drop-down arrow and choose "Same as Layer 'Major Roads.'"

Since *MajorRoads* is in State Plane California Zone V, this becomes the coordinate system environment setting.

Ⓓ Click the Geographic Transformations drop-down arrow.

Ⓔ Scroll down about three quarters of the way and select NAD_1983_To_WGS_1984_5 (Figure 4-14).

The transformation name is added to a box, as shown in Figure 4-15.

Ⓕ Click OK.

Figure 4-13

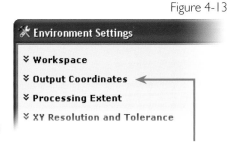

Click here (or on the double arrows) to expand the heading

Figure 4-15

Figure 4-14

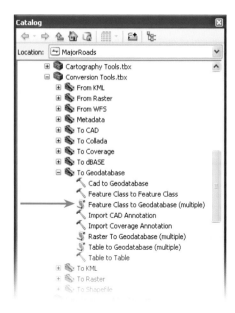

Figure 4-16

By making our desired output coordinate system an environment setting, we tell ArcMap to make sure that whenever a geoprocessing tool runs, the output is projected to that system.

2) Convert multiple feature classes. The *Parkland* feature class is in the WGS 1984 geographic coordinate system and needs to be projected. The *block_centroids* feature class is a shapefile in the WGS 1984 system. It needs to be both converted to geodatabase format and projected.

Ⓐ Scroll down in the Catalog window. Under System Toolboxes, expand Conversion Tools, then expand To Geodatabase.

Ⓑ Right-click the Feature Class to Geodatabase (multiple) tool (Figure 4-16) and choose Open.

▷ Or open the tool by double-clicking it.

Ⓒ In the Catalog window, scroll up to the SourceData folder.

Ⓓ Under the ESRI geodatabase, expand the Landmark feature dataset.

Ⓔ Drag and drop *Parkland* from the Catalog window to the Input Features parameter of the tool.

The feature class is added as the first item in a list underneath the Input Features parameter.

Ⓕ In the Catalog window, expand the census folder.

Ⓖ Drag and drop *block_centroids.shp* to the tool's input parameter.

Ⓗ In the Catalog window, in the AnalysisData folder, click ReadyData.gdb to highlight it.

Ⓘ Drag and drop ReadyData into the tool's Output Geodatabase parameter.

Ⓙ Confirm that your tool dialog box looks like Figure 4-17, then click OK.

The tool runs and informs you when it's done.

3) Inspect the results. Let's see what we've accomplished.

Ⓐ In the Catalog window, expand the ReadyData geodatabase, if necessary.

It should contain three feature classes: *block_centroids*, *MajorRoads*, and *Parkland*.

Ⓑ Check the coordinate systems of the *block_centroids* and *Parkland* feature classes.

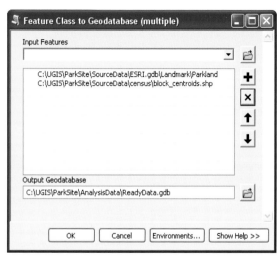

Figure 4-17

This tool projected both input feature classes (and transformed their geographic coordinate systems) thanks to the environment setting you made. It also converted *block_centroids* from shapefile to geodatabase format. As usual with geoprocessing operations, the outputs are new feature classes and the original inputs are unchanged.

You probably noticed that the tool icon 📜 looked like a scroll of paper. Tools with this icon are script tools and they don't always conform to the same protocols as built-in tools 🔨. For example, the tool's output geodatabase parameter didn't default to ReadyData. Nor were the converted feature classes added as layers to the map.

4)   **Rename feature classes.** We'll rename the new feature classes according to our convention.

Ⓐ In the Catalog window, under ReadyData, right-click *block_centroids* and choose Rename.

▷ Or, click on the feature class name to highlight it, then click again to edit it.

The name is surrounded by a black box, making it editable.

Ⓑ Rename it **BlockCentroids** and press Enter.

Ⓒ Rename *Parkland* to **Parks** (Figure 4-18).

This is a good time to save the map document.

Ⓓ On the Standard toolbar, click the Save button 💾.

Ⓔ Navigate, if necessary, to the MapsAndMore folder. Name the map document **Lesson4** and click Save.

In the Catalog window, the home folder changes to ParkSite\MapsAndMore.

Ⓕ If you're continuing to the next exercise now, leave ArcMap open; otherwise, exit ArcMap and close the data requirements table.

Figure 4-18

Tool types

The tools at your disposal depend on your ArcGIS product level (ArcView®, ArcEditor®, or ArcInfo®) and on whether you have extension products installed. Tools have different icons according to their type. Built-in tools 🔨 are written in a compiled programming language and can't be modified. Script tools 📜 are mostly written in the Python® scripting language. You can view, copy, and edit the source code. Many of the script tools that come with ArcGIS adapt built-in tools to work with multiple inputs. Model tools 🔲 encapsulate sequences of data processing operations. In Lesson 7, you'll create a model tool that packages the entire analysis project as a tool.

Exercise 4d: Prepare the city and county data

Now we'll move on to data that needs more preparation. In the requirements table, we listed *City_ply* and *County* as data needed for the final map. We made notes to reduce their extent because both feature classes cover the entire United States.

Processing data to get rid of extra features or attributes is a matter of choice. We could just as well leave the datasets intact and instead apply appropriate definition queries to the *City_ply* and *County* layers in the final map. One point in favor of small datasets is that they generally display and process faster. Likewise, an attribute table is easier to read if it's not cluttered with unnecessary attributes.

We need to be careful in deciding which features and attributes are "unnecessary," because once we get rid of them, they're gone—gone, at least, from that dataset. We'll still have the original source data as a backup, so we can recover from mistakes. We probably wouldn't delete features or attributes that weren't preserved in another feature class.

1)   **Add data to the map.** We'll add the *City_ply* and *County* feature classes as layers to the map. Then we'll select the features we want to keep and copy the selections to new feature classes.

Ⓐ If necessary, start ArcMap and open Lesson4 from the list of recent maps.

Ⓑ In the Catalog window, in the SourceData folder, under ESRI.gdb, expand the Boundary feature dataset.

Ⓒ Click on the Boundary feature dataset name to highlight it.

Ⓓ At the top of the Catalog window, click the Show Next View button 🖫:.

The Catalog window splits into two panes. You can use the bottom pane to drag multiple datasets onto the map.

Ⓔ In the bottom pane, hold down the Control key and click the *City_ply* and *County* feature classes to highlight both (Figure 4-19).

Ⓕ Drag and drop the datasets into the map window.

   ▷ If you get a geographic coordinate systems warning, check the box that says "Don't warn me again in this session" and click Close.

The two layers are added to the map.

Ⓖ In the Table of Contents, click in some empty white space to unselect the layers.

Ⓗ If necessary, drag *City_ply* above *County*.

Figure 4-19

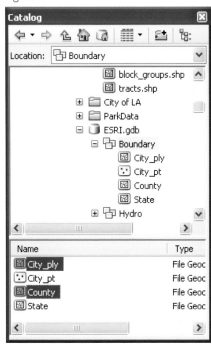

I Optionally, change the symbols to colors that suit you.

J Remove the *MajorRoads* layer from the map.

> ▷ How? In the Table of Contents, right-click the layer name and choose Remove.

2) Select counties. In this step, we'll select the county features that we want to keep.

A Turn off the *City_ply* layer.

B In the Table of Contents, right-click *County* and choose Label Features.

The counties are labeled with their names. Label positions change dynamically with the map scale and extent, so yours may be placed differently.

C If necessary, pan the map to put Los Angeles County roughly in the middle of the display (Figure 4-20).

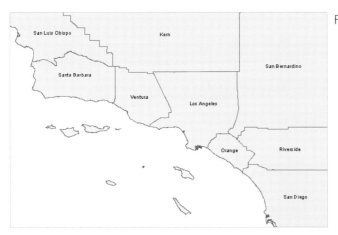

Figure 4-20

D On the Tools toolbar, click the Select Features by Rectangle tool. Click anywhere inside Los Angeles county to select it.

The selected feature is outlined in bright blue (Figure 4-21). This is probably the only county feature we'll need for the final map, but to be safe, we'll select its neighbors as well. We'll do this with a spatial query.

E From the main menu, choose Selection → Select By Location.

Spatial queries evaluate spatial relationships among features. These relationships include proximity, containment, adjacency, and intersection. Typically, features in one layer (the target layer) are selected according to their relationship to features in another layer (the source layer). Sometimes, as here, the target and source layers are the same.

Figure 4-21

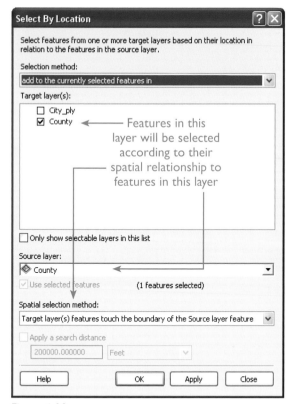

Figure 4-22

ⓕ In the Select By Location dialog box, under Target Layer(s), check the box next to *County*.

ⓖ Click the Source Layer drop-down arrow and choose *County*.

ⓗ At the top of the dialog box, click the Selection method drop-down arrow and choose "add to the currently selected features in."

We want to keep our currently selected feature, Los Angeles County, and select more features.

ⓘ Click the Spatial selection method drop-down arrow. Near the bottom, choose "Target layer(s) features touch the boundary of the Source layer feature."

In plain English, this means that counties bordering Los Angeles County will be selected.

ⓙ Compare your dialog box to Figure 4-22, then click OK.

On the map, Los Angeles, Ventura, Kern, San Bernardino, and Orange counties are selected.

3) Turn off fields. The *County* layer will serve a purely cartographic purpose. That means the only attributes we need are those we might want for labeling or symbology.

ⓐ In the Table of Contents, right-click the *County* layer and choose Open Attribute Table.

ⓑ Scroll quickly across the attributes.

There are about fifty demographic and economic attributes—none of which we need. Removing them will tailor our data to its cartographic purpose.

ⓒ Close the table.

ⓓ Open the layer properties for the *County* layer and click the Fields tab (Figure 4-23).

Figure 4-23

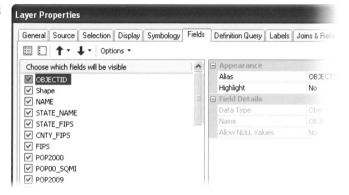

The scrolling box on the left lists the fields in the table. Each field's check box can be unchecked to hide the field. Field visibility is a layer property that you can turn on and off as you like—a hidden field isn't deleted. Nevertheless, the setting does affect the way data is copied. When you copy the feature class, hidden fields will be left out.

E Above the list of fields, click the Turn all fields off button ⬚.

F Click the boxes next to NAME and STATE_NAME to turn just these two fields back on (Figure 4-24). Click OK.

G Open the *County* attribute table to see the effect, then close the table.

H Save the map.

4) Copy the selected features to a new feature class. A geoprocessing tool called Copy Features will copy the five selected features and their attributes to a new feature class. Instead of hunting for the tool among the System toolboxes, we'll make ArcMap get it for us.

A On the right side of the ArcMap application window, click the Search tab to open the Search window.

▷ If the tab isn't showing, click the Search window button 🔍 on the Standard toolbar.

B Near the top of the window, click Tools (Figure 4-25).

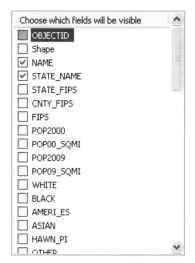

Figure 4-24

Figure 4-25

Figure 4-26

The system toolboxes are displayed in the window.

C In the search box, type **copy**.

As you type, ArcMap presents a list of possible matches in a drop-down box.

D In the drop-down box, click "copy features (data management)" (Figure 4-26).

In the Search window, Copy Features (Data Management) is returned at the top of a list of tools.

E In the list, click on the tool's name, in blue, to open the tool.

▷ If the tool doesn't open, try clicking more quickly.

❻ In the Copy Features dialog box, click the Input features drop-down arrow and choose *County*.

You can choose input features from a drop-down list when there are layers in the map document.

❼ In the Output Feature Class parameter, highlight and delete the entire path. Type **Counties** and press Tab (Figure 4-27).

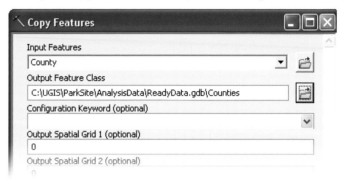

Figure 4-27

❽ Click OK to run the tool.

When the tool is done, a *Counties* feature class is added to ReadyData and a corresponding layer is added to the map.

❾ In the Catalog window, under ReadyData, right-click *Counties* and choose Properties.

❿ In the Feature Class Properties dialog box, click the Fields tab (Figure 4-28).

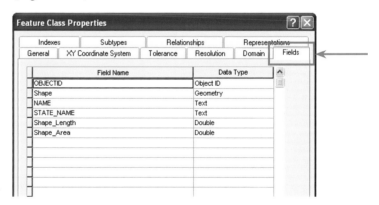

Figure 4-28

The only fields copied from the *County* layer are NAME and STATE_NAME. The other four are created automatically by ArcMap.

Ⓚ Close the Feature Class Properties dialog box.

Ⓛ In the Table of Contents, remove the (old) *County* layer.

Ⓜ Right-click the (new) *Counties* layer and choose Zoom To Layer.

The *Counties* layer consists of the five counties you selected. (The islands are part of the Los Angeles and Ventura county features. Features with discontinuous geometry—islands are a common example—are called "multipart" features. If you click on an island with the Identify tool, you'll see the entire county feature flash green.)

5) Select cities. As with the counties, we don't know exactly which cities we need for the final map. We'll make an interactive selection that takes in an indefinite, but ample, number of features.

Ⓐ In the Table of Contents, turn on the *City_ply* layer.

Ⓑ Drag *City_ply* above *Counties*.

Ⓒ Right-click *City_ply* and choose Selection → Make This The Only Selectable Layer.

Ⓓ On the Tools toolbar, click the Select Features By Rectangle tool 🔲.

Ⓔ Use the tool to drag a selection box similar to the one in Figure 4-29. (Your box can be bigger or smaller.)

When you release the mouse button, cities entirely or partially within the box are selected (Figure 4-30).

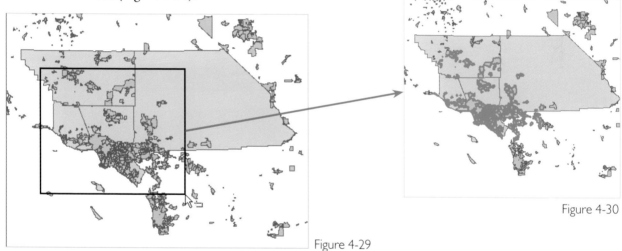

Figure 4-29

Figure 4-30

Ⓕ At the top of the Table of Contents, click the List By Selection button 🔲.

In the example, 298 cities have been selected (Figure 4-31). It doesn't matter if you selected more or fewer.

Ⓖ At the top of the Table of Contents, click the List By Drawing Order button 🔷.

Figure 4-31

Figure 4-32

6) **Turn off fields.** As we did with the counties, we'll get rid of attributes we don't need.

Ⓐ Open the layer properties for the *City_ply* layer and click the Fields tab, if necessary.

Ⓑ Turn off all the fields. 🔲

Ⓒ Click the check boxes next to NAME and ST (an abbreviation for State) to turn these two fields back on (Figure 4-32).

Ⓓ Click OK on the Layer Properties dialog box.

7) **Copy features.** Now we'll copy the selected cities to a new feature class in ReadyData.

Ⓐ Click the Search tab to open the Search window.

▷ Hovering the mouse over the tab will also open the window as long as the ArcMap application window is active.

The Copy Features tool is still at the top of the list.

Ⓑ Click on the tool's name to open it.

Ⓒ In the Copy Features dialog box, click the Input Features drop-down arrow and choose *City_ply*.

Ⓓ In the Output Feature Class parameter, clear the path. Type **Cities** and press Tab.

Ⓔ Compare your tool dialog box to Figure 4-33, then click OK.

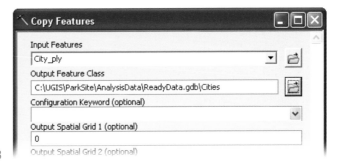

Figure 4-33

When the tool is finished, a *Cities* feature class is created in ReadyData and a layer is added to the map.

Ⓕ In the Catalog window, confirm that you have a *Cities* feature class in ReadyData.

Ⓖ In the Table of Contents, remove the (old) *City_ply* layer.

Ⓗ Open the attribute table for the (new) *Cities* layer.

The table contains the two attributes you kept, plus the software-managed attributes (Figure 4-34).

Figure 4-34

- Close the table.
- Save the map document ![disk icon].
- If you're continuing to the next exercise now, leave ArcMap open; otherwise, exit ArcMap.

Using the Search window

The Search window lets you search for map documents and data as well as tools. Tools are indexed automatically by the software, but to search for maps and data, you have to index the workspaces that you want searched. To create an index, click the Index / Search Options button ![icon] at the top of the Search window. In any search, you can type keywords into the search box to help you find items that you don't know by name. Keyword searches use tags in the item's metadata.

Exercise 4e: Prepare the river data

In Row 5 of the data requirements table, we made a note to dissolve the *LARiver* dataset. This will combine the river's 265 features into one, simplifying the task of creating a buffer around the river in Lesson 6. The Dissolve tool creates a new feature class. When we run the tool, we'll convert the *LARiver* shapefile to a geodatabase feature class as part of the operation.

1) **Dissolve the river.** A few common geoprocessing tools, including Dissolve, can be accessed from the main menu.

- If necessary, start ArcMap and open Lesson4 from the list of recent maps.
- From the main menu, choose Geoprocessing → Dissolve.
- In the Catalog window, under SourceData, expand the City of LA folder.
- Drag *LARiver.shp* to the tool's Input Features parameter.

The output feature class location defaults to ReadyData.

Dissolving features

You can dissolve a feature class either by geometry or by an attribute. If you dissolve by geometry, common boundaries between features disintegrate. The number of output features depends on whether the "Create multipart features" box is checked. If it's checked (the default), you'll get one output feature with discontinuous geometry. If you uncheck it, you'll get a unique, singlepart feature for each spatially distinct area. In either case, the output table has no attributes except for those maintained by ArcMap.

☑ Create multipart features (optional)　　　☐ Create multipart features (optional)

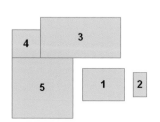
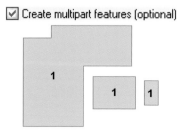
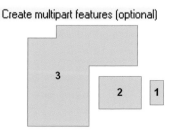

Five input features will be dissolved by geometry

The output is one multipart feature

The output is three singlepart features

If you dissolve by an attribute, common boundaries between features disintegrate only if the features have the same attribute value. Again, the number of output features depends on whether multipart features are created. If so, all input features with the same attribute value will become one output feature (with discontinuous geometry). If not, you'll get a unique feature for each spatially distinct area. The attribute being dissolved on is preserved in the output table.

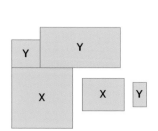
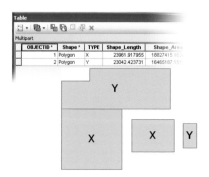
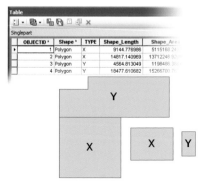

Five input features will be dissolved on a TYPE attribute

The output is two multipart features: one X feature and one Y feature

The output is four singlepart features: two X features and two Y features

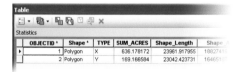

In addition to the attribute being dissolved on, you can set a "statistics field" to summarize the values of dissolved features for another attribute. In this example, the features are dissolved by TYPE and SUM_ACRES is a statistics field (based on an ACRES field in the original table) that calculates the total acreage of the dissolved features.

E Change the output feature class name from LARiver_Dissolve to **LARiver**.

▷ You can use the method of clearing the path and typing the new name, or just backspace over the part of the name you want to remove.

F In the Dissolve Fields area, check the NAME box.

Checking this box preserves the NAME field, which we want to keep, in the output attribute table. It also affects the way the tool works. Instead of creating a single output feature, ArcMap will create a feature for each unique value in the checked field. In our case, it doesn't change the output because the NAME field only has one value ("Los Angeles River"). In other situations—for instance, if our feature class had many different river names—it would consolidate all features with the same name into single features.

G Compare your dialog box to Figure 4-35, then click OK.

When the tool is done, a feature class is added to ReadyData and a layer is added to the map.

H Open the attribute table for the *LARiver* layer (Figure 4-36).

It contains one record. Except for the NAME field you checked, the attributes from the original table have been dropped.

I Close the attribute table.

J Optionally, symbolize the river in a shade of blue on the map.

Figure 4-35

Exercise 4f: Prepare the park data

The *Parks* feature class is already in the geodatabase, but we still have the preparation task of loading data into it. That's because this dataset is missing the two park features that we saw in the *NewParks* shapefile.

1) Add data to the map. Let's remind ourselves which two parks we're talking about.

A From the ReadyData geodatabase, add *Parks* to the map. Optionally, symbolize the layer in a shade of green.

B Under SourceData, expand the ParkData folder and add *NewParks.shp*. Optionally, symbolize this layer in a different shade of green.

Figure 4-36

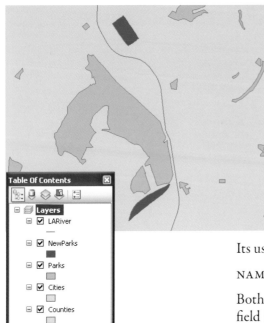

Figure 4-37

C Zoom to the *NewParks* layer (Figure 4-37).

2) Compare attributes. We want to get these two *NewParks* features, along with their attributes, into the *Parks* feature class. We face a complication in that the datasets have different attributes.

A Open the *Parks* attribute table.

It has 1,209 records. Its user-managed attributes are:

NAME   TYPE   ACRES

The TYPE attribute describes whether the park is a local, state, or national park.

B Open the *NewParks* attribute table.

Its user-managed attributes are:

NAME   CATEGORY   ADDRESS   TELEPHONE   HOURS

Both tables have a NAME field. CATEGORY is similar to the TYPE field in the *Parks* table in that it identifies the parks as state parks. The other fields don't have similar fields in the *Parks* table.

C Close the table window.

3) Load data. We'll load the data and see how ArcMap handles the attribute mismatches.

A In the Catalog window, under ReadyData, right-click *Parks* and choose Load → Load Data.

The Simple Data Loader wizard opens.

B On the wizard panel, click Next to advance.

C Next to Input data, click the Open Folder button.

D In the Open GeoDatabase dialog box, browse to the SourceData\ParkData folder. Click *NewParks.shp*, and click Open.

E On the wizard panel, click the Add button.

NewParks.shp is the data we want to load (Figure 4-38).

F Click Next to advance.

This panel confirms that the data will be loaded into an existing geodatabase (ReadyData.gdb) and into the feature class *Parks*. Neither setting can be changed.

G Click Next to advance.

Figure 4-38

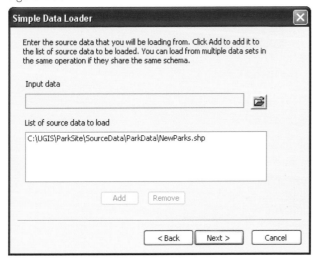

In the next panel, you specify how to handle attributes. ArcMap assumes that you want NAME values from the source feature class (*NewParks*) to be copied into the NAME field of the target feature class (*Parks*). It doesn't make any other assumptions because the other field names don't match (Figure 4-39).

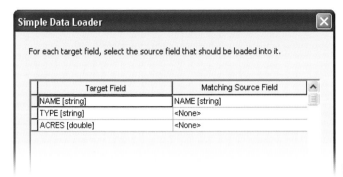

Figure 4-39

❽ In the second row (where TYPE is the Target Field value), click on <None> in the Matching Source Field column.

A drop-down list appears with the attributes from *NewParks*.

❾ In the list of attributes, click CATEGORY (Figure 4-40).

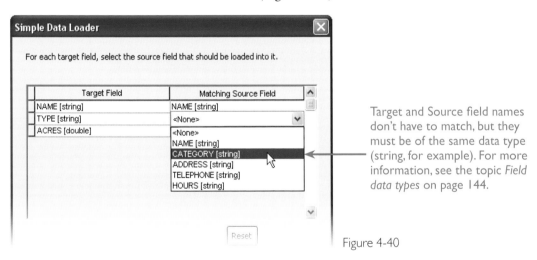

Target and Source field names don't have to match, but they must be of the same data type (string, for example). For more information, see the topic *Field data types* on page 144.

Figure 4-40

Values from the CATEGORY field in *NewParks* will be copied into the TYPE field of *Parks*.

There's no ACRES field (by this or any other name) in the *NewParks* table, so leave the Matching Source Field in the third row set to <None>. No values will be copied into the ACRES field in *Parks*.

The ADDRESS, TELEPHONE, and HOURS attributes in *NewParks* simply won't be loaded, because *Parks* has no appropriate attributes to accept the values.

Field data types

One of the essential properties of a field in an attribute table is its field data type, which defines the type of information the field can hold. When a new field is added to a table, its field data type is specified and can't be changed thereafter. There are several field data types, but those used most often are text (sometimes called string) and the four numeric types: short integers, long integers, floats, and doubles.

Text fields are for descriptions, codes, noncomputational numbers (like postal codes or telephone numbers), and the like. The default length of a text field is 50 characters, which means that 50 bytes of space are reserved for each entry. If you know that your entries will be shorter—for example, if you're storing two-letter state abbreviations—you should set the length property accordingly. In the interest of saving space, you can also use codes in place of long descriptions. In a geo-database, codes can be stored, managed, and enforced through a so-called attribute domain (not discussed in this book).

For numeric data, use short or long integers when the values are whole numbers, such as population or number of items sold. Whether to use the short or long integer type depends on how large the numbers may be (see the table below). Use floats or doubles when the values are fractional. Floats store six digits with precision (for example, 12345.6 or 1.23456). Numbers with more digits are stored with some rounding, which may affect calculations slightly. Doubles are stored with fifteen digits of precision. Floats are suitable for most statistical calculations, such as population density or average income. Doubles are recommended for fields storing geographic measurements and calculations. ArcMap uses the Double type in its Shape_Length and Shape_Area fields.

The following table summarizes the field data types available for geodatabase feature classes. Not all are available for shapefiles.

Field Type	Stores
Text	Letters, numbers, special characters
Short integer	Whole numbers from -32,768 to 32,767 (uses 2 bytes)
Long integer	Whole numbers from -2,147,483,648 to 2,147,483,647 (uses 4 bytes)
Float	Fractional numbers: precise to 6 digits, then rounded (uses 4 bytes)
Double	Fractional numbers: precise to 15 digits, then rounded (uses 8 bytes)
Date	Dates and/or times
BLOB	Binary large objects, including special feature types like geodatabase annotation
Raster	Small images
GUID	Unique feature/record identifier for features in distributed geodatabases
ObjectID	Unique feature/record identifer
Shape	Feature geometry type (for example, point, line, or polygon)

J Confirm that the wizard panel looks like Figure 4-41.

  ▷ If you made a mistake, click Reset on the wizard panel and start
    again.

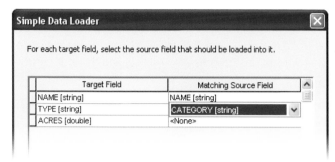

Figure 4-41

K Click Next to advance.

The default setting on this panel is to load all the source data. That's
what we want.

L Click Next to advance.

M Click Finish on the summary panel.

When you load data, you don't get any messages. The operation
happens quickly and invisibly: it's already done.

4) View results. Let's see what we accomplished.

A In the Table of Contents, turn off the *NewParks* layer.

The new features now show up in the *Parks* layer (Figure 4-42).

B Open the Parks attribute table and scroll to the bottom.

The table now has 1,211 records, instead of the 1,209 it had before.
The last two records are the ones that were loaded. They have values
in the NAME and TYPE fields, but not in the ACRES field (Figure
4-43).

C Close the table.

D Remove the *NewParks* layer from the Table of Contents.

Figure 4-42

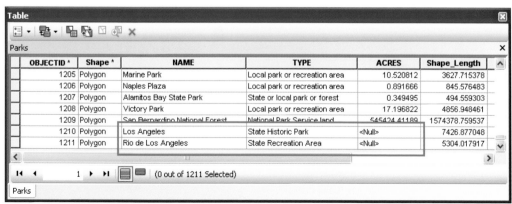

OBJECTID *	Shape *	NAME	TYPE	ACRES	Shape_Length
1205	Polygon	Marine Park	Local park or recreation area	10.520812	3627.715378
1206	Polygon	Naples Plaza	Local park or recreation area	0.891666	845.576483
1207	Polygon	Alamitos Bay State Park	State or local park or forest	0.349495	494.559303
1208	Polygon	Victory Park	Local park or recreation area	17.196822	4856.948461
1209	Polygon	San Bernardino National Forest	National Park Service land	545424.41189	1574378.759537
1210	Polygon	Los Angeles	State Historic Park	<Null>	7426.877048
1211	Polygon	Rio de Los Angeles	State Recreation Area	<Null>	5304.017917

(0 out of 1211 Selected)

Parks

Figure 4-43

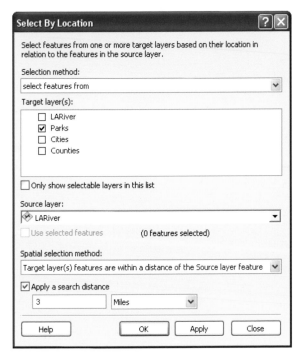

Figure 4-44

5) Make a spatial selection on the *Parks* layer. It's not one of our preparation tasks to reduce the *Parks* data, but the spatial extent of this feature class is bigger than it needs to be. Getting rid of extra features may help speed up some analysis operations in Lessons 6 and 7.

Ⓐ Zoom to the *Parks* layer.

Ⓑ Turn off the *Cities* layer.

Those big national forests to the north are definitely outside our area of interest.

Ⓒ From the ArcMap main menu, choose Selection → Select By Location.

Ⓓ In the Select By Location dialog box, set the selection method, if necessary, to "select features from."

Ⓔ Under Target Layer(s), check *Parks*.

Ⓕ Click the Source Layer drop-down arrow and choose *LARiver*.

Ⓖ Click the Spatial selection method drop-down arrow and choose "Target layer(s) features are within a distance of the Source layer feature" (the third entry in the list).

Ⓗ Highlight and delete whatever numeric value is in the box underneath "Apply a search distance." Type **3**.

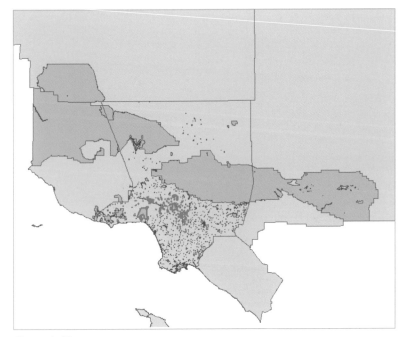

Figure 4-45

Ⓘ Click the drop-down arrow in the adjacent box and set the units to Miles.

The spatial query will select parks within three miles of the river. If any bit of a park is within this distance, the feature will be selected. All the unselected features will be well out of our area of interest.

Ⓙ Compare your dialog box to Figure 4-44, then click OK.

On the map, park features within three miles of the river are selected (Figure 4-45).

Ⓚ At the top of the Table of Contents, click the List By Selection button 🔳.

There should be 169 selected parks (Figure 4-46).

Ⓛ Click the List By Drawing Order button 🗂.

6) **Copy features.** Now we'll copy the selected parks, but we have to create a new output feature class. The *Parks* layer, with its 169 selected features, already references the *Parks* feature class in ReadyData, and ArcMap won't let it overwrite itself.

Ⓐ In the Search window, locate and open the Copy Features tool.

Ⓑ In the tool dialog box, set the Input Features to the *Parks* layer.

Ⓒ Clear the Output Feature Class path, type `Parks2`, and press Tab.

Ⓓ Compare your Copy Features dialog box to Figure 4-47, then click OK.

Figure 4-46

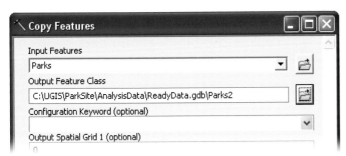

Figure 4-47

When the tool is finished, *Parks2* is created in ReadyData and a layer is added to the map.

Ⓔ In the Table of Contents, turn off the (old) *Parks* layer.

Ⓕ Resymbolize *Parks2* in any shade of green.

Ⓖ Zoom to the *Parks2* layer and open its attribute table.

There should be 169 records.

Ⓗ Close the table.

We don't need the *Parks* feature class any more, so we'll delete it from the geodatabase.

Ⓘ In the Catalog window, under ReadyData, right-click *Parks* and choose Delete.

Make sure *Parks*—not *Parks2*—is highlighted in the Catalog window.

Ⓙ Click Yes on the prompt to confirm your deletion.

The feature class is deleted from ReadyData and the corresponding layer is removed from the map. You still have the original *Parkland* data in your SourceData folder so you can recover from mistakes if you have to.

ⓚ In the Catalog window, right-click *Parks2* and choose Rename. Rename the feature class to *Parks* and press Enter.

The map layer pointing to this feature class is still called *Parks2,* which is fine.

ⓛ Save the map document.

ⓜ If you're continuing to the next exercise now, leave ArcMap open; otherwise, exit ArcMap.

Overwriting geoprocessing outputs

By default, ArcMap doesn't allow geoprocessing outputs to overwrite existing datasets. This is a useful safeguard, but if you need to change it, choose Geoprocessing Options from the Geoprocessing menu and check the box to "Overwrite the outputs of geoprocessing operations." (Changing this setting still won't allow you to overwrite a feature class with itself.)

Exercise 4g: Prepare the block group data

In Rows 8 and 9 of the data requirements table we identified some block group attributes—population density and age under 18—that we need to calculate from our existing data. We'll add new fields to the table, then calculate values into them.

1) **Add data to the map.** First, we'll add the *block_groups* shapefile to the map and have another look at it before we copy it to ReadyData.

ⓐ If necessary, start ArcMap and open Lesson4.

ⓑ In the Catalog window, under SourceData, expand the census folder. Drag and drop *block_groups.shp* onto the map.

  ▷ If you get a Geographic Coordinate Systems Warning, close it.

ⓒ Zoom to the *block_groups* layer (Figure 4-48).

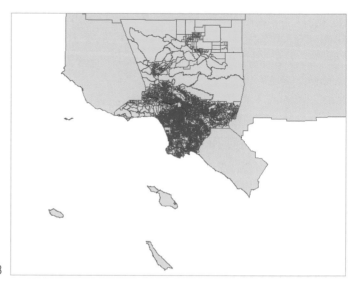

Figure 4-48

The spatial extent of the layer is Los Angeles County.

○ Open the *block_groups* attribute table and scroll across the attributes.

We only need a handful of the twenty or so attributes in the table. As with the parks, we didn't make a note to reduce this dataset, but it will be easier to work with the important attributes if we don't have to keep scrolling past the unimportant ones.

○ Close the table.

2) **Turn off fields.** We'll turn off the fields we don't need.

○ Open the layer properties for the *block_groups* layer and click the Fields tab, if necessary.

○ Click the Turn all fields off button ▤ .

○ Check the boxes for the following fields to turn them back on:

- ID (a unique user-managed identifier)
- ST_ABBREV (a two-letter designation for the state)
- TOTPOP_CY (total population)
- POP18UP_CY (population over 18)
- MEDHINC_CY (median household income)

○ Confirm that your list of visible fields matches Figure 4-49. (You'll need to scroll down in your list.)

○ Click OK.

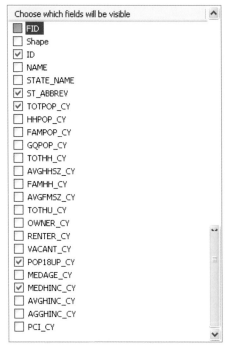

Figure 4-49

3) **Import block groups.** Earlier, we used the Feature Class to Geodatabase (multiple) tool to get *BlockCentroids* and *Parks* into ReadyData. This time we'll use Feature Class to Feature Class, a similar tool that converts one feature class at a time. It has some advantages over Feature Class to Geodatabase (multiple), as well as over Copy Features, in that it lets you modify the properties of fields in the output table. The Feature Class to Feature Class tool is stored in the Conversion Tools toolbox, but it can also be opened from a geodatabase's context menu.

○ In the Catalog window, right-click the ReadyData geodatabase and choose Import → Feature Class (single).

The Feature Class to Feature Class tool opens.

○ In the Input Features parameter, click the drop-down arrow and choose *block_groups*.

The output location is automatically set to ReadyData.

○ In the Output Feature Class parameter, type **BlockGroups**.

○ In the Field Map area, right-click the ST_ABBREV field and choose Properties.

ⓔ In the Output Field Properties dialog box, change the name to **STATE**. Change the alias to **STATE** as well.

For more information about aliases, see the box on page 152.

ⓕ Compare your Output Field Properties dialog box to Figure 4-50, then click OK.

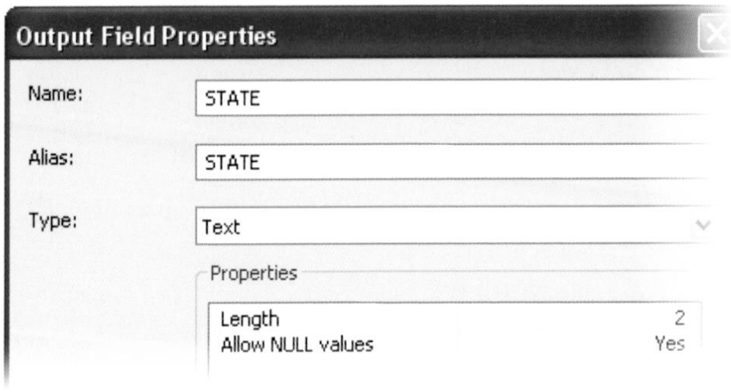

Figure 4-50

ⓖ In the Field Map area of the tool, right-click the TOTPOP_CY field and choose Properties.

ⓗ Change the name to **TOTPOP** and change the alias to **TOTAL POP** (with a space).

ⓘ Click the Type drop-down arrow and change the type from Long to Short (Figure 4-51). Click OK.

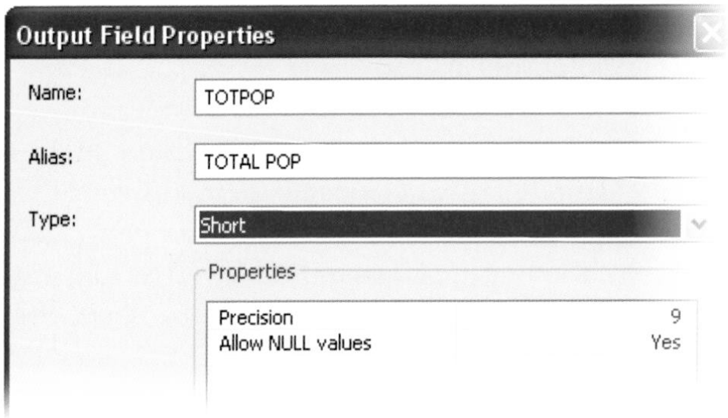

Figure 4-51

Every field has a field data type that specifies what kind of information it holds. The Long and Short types both store integers. The largest value that can be stored by the Short type is 32,767. Since no block groups have a population that big, this type is adequate for our needs. See the topic *Field data types* on page 144 for more information.

J Change the output field properties for the remaining fields as follows:

- For POP18UP_CY, change the name to **POP18UP** and the alias to **POP OVER 18.** Change the field type from Long to Short.
- For MEDHINC_CY, change the name to **MEDHINC** and the alias to **MEDIAN HH INCOME.** (Don't change the field type.)

K Compare your Feature Class to Feature Class dialog box to Figure 4-52, then click OK to run the tool.

When the tool is finished, a new feature class is created in ReadyData and a layer is added to the map.

L In the Table of Contents, remove the (old) *block_groups* layer.

M Open the attribute table for the (new) *BlockGroups* layer.

The table contains the attributes you kept and displays their aliases (Figure 4-53).

Figure 4-52

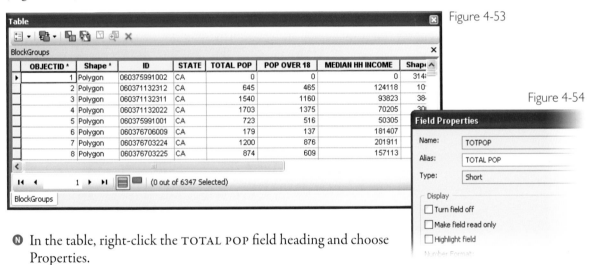

Figure 4-53

Figure 4-54

N In the table, right-click the TOTAL POP field heading and choose Properties.

In the Field Properties dialog box, the field name, alias, and type are all set as specified (Figure 4-54).

O Close the Field Properties dialog box, but leave the table open.

Field names and aliases

Field names in a geodatabase table are limited to 31 characters. They must begin with a letter and cannot include spaces. A few reserved words (for example, "Add," "Drop," and "Select") can't be used as field names. If you try to name a field with a disallowed name, you'll be prompted to rename the field. Field names in a shapefile table are limited to 10 characters. They have the same naming restrictions as geodatabase tables, except that there are no reserved words. Once set, the name of a field can't be changed.

An alias is a flexible and descriptive alternative name that can be displayed instead of a field's actual name. For fields in geodatabase feature classes, an alias can be set either as a feature class property or as a layer property. If it's a feature class property, it will be used whenever the feature class is added as a layer. If it's a layer property, it will be used only for the particular layer in which it is set. For fields in shapefiles, aliases are supported only as layer properties.

4) Calculate population density. Population density is the ratio of population to area. The SHAPE_AREA field stores the area of each block group in square feet. (This field was added automatically by ArcMap when we converted the block groups to geodatabase format.) We know the units are square feet because those are the measurement units specified by our coordinate system. Since we need to express density in terms of people per square mile, we're going to add two fields to the *BlockGroups* attribute table: one to store area in square miles and another to store population density expressed in square miles. Then we'll make the calculations for both fields.

Ⓐ At the top of the table, click the Table Options button 🔳. On the context menu, choose Add Field.

Ⓑ In the Add Field dialog box, for the name, type **SQMILES**.

Ⓒ For Type, click the drop-down arrow and choose Double.

Ⓓ In the Field Properties area, click in the white box next to Alias and type **SQ MILES** (Figure 4-55). Click OK.

Ⓔ Scroll to the end of the table to see the new field.

It has <Null> values for all records. For more information, see the box *Null values* on page 158.

Ⓕ Add another field and give it the following properties:
  - Name: **POPDENSITY**
  - Type: Float
  - Alias: **POP DENSITY**

Ⓖ Compare your Add Field dialog box to Figure 4-56, then click OK.

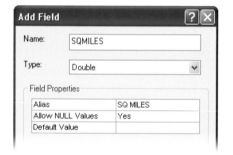

Figure 4-55

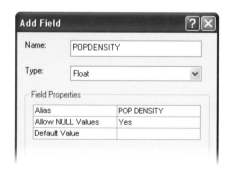

Figure 4-56

Usually, when you make changes to a layer in ArcMap, you affect only the layer and not the feature class it points to. Changing the structure of a table by adding or deleting fields is an exception.

H In the Catalog window, under ReadyData, right-click *BlockGroups* and choose Properties.

I In the Feature Class Properties dialog box, click the Fields tab.

You can see that both fields have been added to the feature class.

J At the bottom of the list of fields, click the gray box to the left of the POPDENSITY field name (Figure 4-57).

Click here to see the field properties.

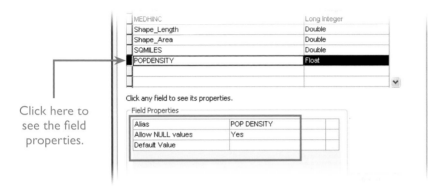

Figure 4-57

The alias that you typed (POP DENSITY) has been added, too.

K Close the Feature Class Properties dialog box.

L In the *BlockGroups* attribute table, right-click the SQ MILES field heading and choose Calculate Geometry.

A warning message appears (Figure 4-58).

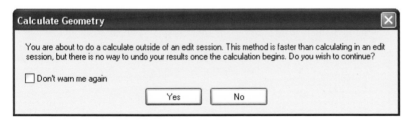

Figure 4-58

You can calculate and edit attribute values without starting an "edit session"—we're about to do it—but you give up the option of undoing a mistake. That's a problem if you overwrite data that you can't recover, but in many situations you can fix a mistake by running the calculation again.

M On the Calculate Geometry warning, check the box that says "Don't warn me again," then click Yes.

N In the Calculate Geometry dialog box, make sure the Property is Area and the coordinate system option is "Use coordinate system of the data source."

O Click the Units drop-down arrow and choose Square Miles US [sq mi] (Figure 4-59). Click OK.

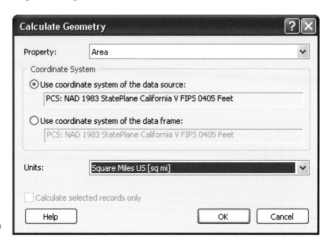

Figure 4-59

ArcMap calculates the area in square miles for each record and adds the values to the table (Figure 4-60).

POP 18 UP	MEDIAN HH INCOME	Shape_Length	Shape_Area	SQ MILES	POP DENSITY
0	0	314839.404519	1597144137.48898	57.289663	<Null>
465	124118	10123.303831	5918525.55489B	0.212298	<Null>
1160	93823	38414.772596	42606492.03508B	1.528298	<Null>
1375	70205	30000.068439	18234928.4473B	0.654088	<Null>
516	50305	337601.826827	2064229455.6317B	74.044043	<Null>
137	181407	12611.677293	7703066.20207B	0.276309	<Null>
876	201911	27705.95831	12135059.00509B	0.435285	<Null>
609	157113	12731.538805	6934825.3063B	0.248753	<Null>

Figure 4-60

P In the table, right-click the POP DENSITY field heading and choose Field Calculator.

In the Field Calculator dialog box, ArcMap automatically supplies the first part of the expression: "POPDENSITY = ."

Q In the list of fields, double-click TOTPOP to add it to the expression box.

R Click the Division button ⟦ / ⟧ .

⑤ In the list of fields, double-click SQMILES.

The complete expression should be:

POPDENSITY = [TOTPOP] / [SQMILES] (Figure 4-61).

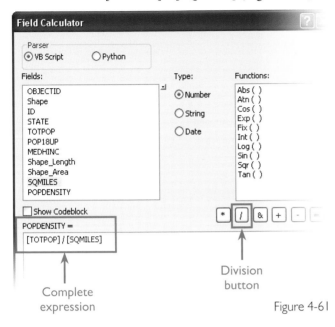

Complete expression

Division button

Figure 4-61

⑥ Click OK.

ArcMap calculates the population density for each record and adds the values to the table (Figure 4-62).

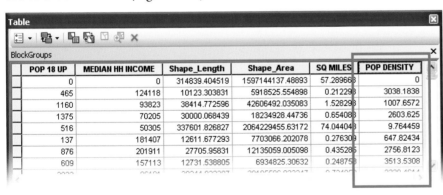

POP 18 UP	MEDIAN HH INCOME	Shape_Length	Shape_Area	SQ MILES	POP DENSITY
0	0	314839.404519	1597144137.48893	57.289668	0
465	124118	10123.303831	5918525.554898	0.212298	3038.1838
1160	93823	38414.772596	42606492.035083	1.528298	1007.6572
1375	70205	30000.068439	18234928.44736	0.654088	2603.625
516	50305	337601.826827	2064229455.63172	74.044048	9.764459
137	181407	12611.677293	7703066.202078	0.276309	647.82434
876	201911	27705.95831	12135059.005098	0.435286	2756.8123
609	157113	12731.538805	6934825.30632	0.248758	3513.5308

Figure 4-62

5) Add fields to calculate the percentage under 18. To get the percentage of the population under age 18, we'll subtract the 18-and-over population from the total population, then divide the result (the under-18 population) by the total population. We'll then multiply this value by 100 to express our percentage as a whole number. Again, we'll need two new fields to hold our calculations.

Ⓐ At the top of the table, click the Table Options button and choose Add Field.

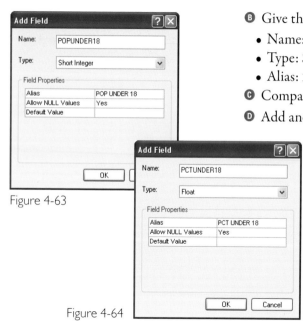

Figure 4-63

Figure 4-64

B Give the new field these properties:
- Name: POPUNDER18
- Type: Short Integer
- Alias: POP UNDER 18

C Compare the Add Field dialog box to Figure 4-63. Click OK.

D Add another field with these properties::
- Name: PCTUNDER18
- Type: Float
- Alias: PCT UNDER 18

E Compare your Add Field dialog box to Figure 4-64, then click OK.

▷ If you realize, after adding a field, that you've made a mistake—especially if you've set the wrong field data type—right-click the field heading in the table, choose Delete Field, and repeat the operation.

The two new fields are added to the end of the table with <Null> values.

6) Calculate the fields. Now we'll calculate values for the fields.

A In the table, right-click the POP UNDER 18 field heading and choose Field Calculator.

B Click Clear to remove the previous expression.

C In the list of fields, double-click TOTPOP to add it to the expression box.

D Click the Minus button ⊡.

E In the list of fields, double-click POP18UP.

The finished expression should be:

POPUNDER18 = [TOTPOP] - [POP18UP] (Figure 4-65).

F Click OK on the Field Calculator dialog box.

ArcMap calculates the population under age 18 for each record (Figure 4-66).

Figure 4-65

Figure 4-66

Shape_Length	Shape_Area	SQ MILES	POP DENSITY	POP UNDER 18	PCT UNDER 18
314839.404519	1597144137.48893	57.289663	0	0	<Null>
10123.303831	5918525.554898	0.212298	3038.1838	180	<Null>
38414.772596	42606492.035083	1.528298	1007.6572	380	<Null>
30000.068439	18234928.44736	0.654088	2603.625	328	<Null>
337601.826827	2064229455.63172	74.044043	9.764459	207	<Null>
12611.677293	7703066.202078	0.276309	647.82434	42	<Null>
27705.95831	12135059.005098	0.435285	2756.8128	324	<Null>
12731.538805	6934825.30632	0.248753	3513.5308	265	<Null>

The last block group calculation we have to make is to divide the under-18 population by the total population to get the percentage of children. We have a potential problem with records that have a total population of 0.

G Scroll back across the table to see the TOTAL POP field.

The first record is a case in point. No one lives in that block group. Division by zero is undefined, and we'll get an error message if we try it. Instead, we'll use an attribute query to select the records with nonzero values in the TOTAL POP field. Then we'll run the calculation on just those records.

H At the top of the table, click the Select by Attributes button ▦.

I In the Select by Attributes dialog box, double-click "TOTPOP" in the list of fields to add it to the expression box.

J Go on to make the following expression: "TOTPOP" > 0.

By default, this dialog box, like the Field Calculator, displays field names, not aliases. (If you want, you can change this setting with the tiny drop-down arrow to the right of the field names box.)

K Compare your expression to Figure 4-67, then click Apply.

L Close the Select by Attributes dialog box.

In the table, 6,316 of 6,347 records are selected.

M Scroll to the end of the table, right-click the PCT UNDER 18 field heading, and choose Field Calculator.

N Click Clear to remove the previous expression.

O In the expression box, make the following expression: ([POPUNDER18] / [TOTPOP]) * 100.

▷ Make sure to include the parentheses as shown.

P Compare your expression to Figure 4-68, then click OK.

Values are calculated for the selected records. The unselected records still have <Null> values. We'll calculate these to zero, because if no one lives in the block group, it's reasonable to say that the percentage of children is zero.

Q At the top of the table window, click the Switch Selection button ▦.

A message warns you that the operation might take a long time, but don't worry—it won't.

R Click Yes to continue.

S Right-click the PCT UNDER 18 field heading again and choose Field Calculator. Clear the previous expression.

T In the expression box, type 0, then click OK.

U At the top of the table, click the Clear Selection button ▦.

Figure 4-67

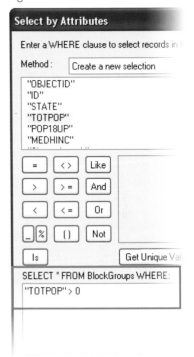

Figure 4-68

Type the parentheses from the keyboard. Spaces between parentheses and field name brackets are optional.

We now have the attributes we need for population density and percentage of the population under age 18 (Figure 4-69).

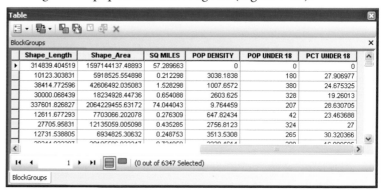

Shape_Length	Shape_Area	SQ MILES	POP DENSITY	POP UNDER 18	PCT UNDER 18
314839.404519	1597144137.48893	57.289663	0	0	0
10123.303831	5918525.554898	0.212298	3038.1838	180	27.906977
38414.772596	42606492.035083	1.528298	1007.6572	380	24.675325
30000.068439	18234928.44736	0.654088	2603.625	328	19.26013
337601.826827	2064229455.63172	74.044043	9.764459	207	28.630705
12611.677293	7703066.202078	0.276309	647.82434	42	23.463688
27705.95831	12135059.005098	0.435285	2756.8123	324	27
12731.538805	6934825.30632	0.248753	3513.5308	265	30.320366

(0 out of 6347 Selected)

Figure 4-69

V Close the table window.

W Save the map.

X If you're continuing to the next exercise now, leave ArcMap open; otherwise, exit ArcMap.

Null values

When a field is added to a geodatabase table, it's populated with <Null> values. A <Null> value indicates the absence of data and usually means that data hasn't been obtained or that values are unknown for some reason. <Null> values are different from zeroes and blank text strings. The query "FIELDNAME" IS NULL will select records with <Null> values. It will not select records with blank text strings or zero values.

An important difference between the geodatabase feature class and shapefile formats is that shapefiles don't support <Null> values. When you add a text field to a shapefile, it's populated with blank values. When you add a numeric field, it's populated with zeroes. If you convert the shapefile to a geodatabase feature class, the blank strings and zeroes will be preserved, which may cause confusion. In shapefiles, arbitrary codes, such as -9999, are often used to represent a <Null> value.

Exercise 4h: Prepare the parcel data

We have to join the stand-alone *Vacant Parcels* table to the *Parcels* feature class table (Row 2 of the data requirements table). We need the attribute of vacancy in the *Parcels* table so that we can find and select vacant parcels on the map. A table join attaches the attributes of one table (usually, a nonspatial stand-alone table) to those of another (usually, a feature class attribute table). The join is based on a field of values, typically an identification code, that is common to both tables. Values in the common field are used to match records in the stand-alone table with records in the feature attribute table. For more information, see the topic *Table joins and relates* on page 160.

1) **Add data.** First, we'll add *Parcels* and *Vacant Parcels* to the map document.

Ⓐ If necessary, start ArcMap and open Lesson4 from the list of recent maps.

Ⓑ In the Table of Contents, turn off all the layers except *LARiver* and *Counties* to simplify the map display.

Ⓒ In the Catalog window, in the SourceData folder, under City of LA, add *Parcels.shp* and *Vacant Parcels.dbf* to the map.

In the Table of Contents, the view changes to List By Source ☟ because of the stand-alone table.

Ⓓ Zoom to the *Parcels* layer.

2) **Identify the common attribute.** We'll look at both tables to find the common attribute, or key field, to join on. Actually, when we viewed the *Vacant Parcels* item description in Lesson 2, we saw that the PIN field serves this purpose, but we should confirm that for ourselves.

Ⓐ Open the *Parcels* attribute table, then the *Vacant Parcels* table.

Ⓑ At the bottom of the table window, drag and drop the Vacant Parcels tab onto the blue arrow pointing to the right (Figure 4-70).

▷ You don't see the arrows until you start to drag the tab.

Ⓒ In the *Parcels* table (on the left side of the table window), scroll all the way to the right to see the PIN attribute.

▷ Optionally, enlarge the table window by dragging a side or corner.

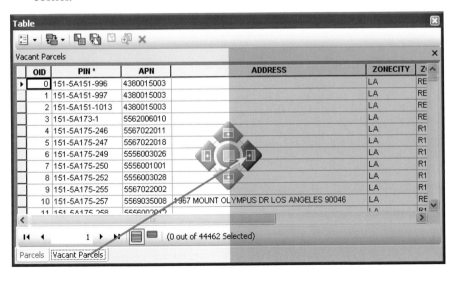

To see both tables at the same time, drag a table by its tab and drop it on a blue arrow.

Figure 4-70

Table joins and relates

Tables can be associated in ArcMap through either a join or a relate. A join attaches one table's attributes to the other table. (The join is "virtual," meaning that it exists only in the map document; the two tables remain separate on disk.) A relate is a look-up relationship, in which you select records in one table to see matching records in the other table. A join is stronger than a relate because attributes in a joined table can be used to symbolize and query features on the map, while attributes from a related table cannot. Both joins and relates require a common attribute to match records.

A join is appropriate when, for any given record in the feature class table (the table *to* which attributes are being attached), there is at most one possible match in the stand-alone table (the table *from* which attributes are being attached). A relate is appropriate when more than one match is possible. Consider some examples:

1) A feature class of parcels and a stand-alone table of vacant parcel attributes, where the common attribute is PARCEL ID, a unique identifier in both tables. Each record in the *Parcels* table has at most one matching record in the *Vacant Parcels* table, so a join may be used. The records have a one-to-one relationship: any record in either table has only one match in the other table.

2) A feature class of rivers and a stand-alone table of watershed attributes, where the common attribute is WATERSHED ID. This is a nonunique identifier for rivers, because all rivers that drain into the same watershed, or drainage area, have the same WATERSHED ID. It is a unique identifier for watersheds. Because a river drains into just one watershed, each record in the *Rivers* table has at most one matching record in the *Watersheds* table, which means that a join may be used. The records have a many-to-one relationship: many records in the feature class table can match the same record in the stand-alone table.

The same watershed record, such as Los Angeles River Watershed, may match many river features.

Attribute from *Rivers* feature class   Key field   Attribute from *Watersheds* table

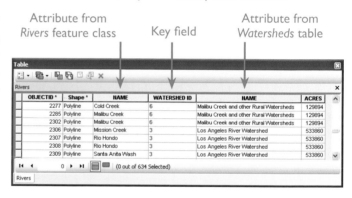

3) A feature class of the Los Angeles River and a stand-alone table of improvement project attributes, where the common attribute is RIVER ID. This is a unique identifier for river features. It is a nonunique identifier for the stand-alone table, which may have more than one improvement project located along the same river feature. In this situation, a relate should be used because a given record in the feature class table may have more than one match in the stand-alone table. In a join, only one matching record could be attached to the feature class table; the others would be left out. The records have a one-to-many relationship: one record in the feature class table may have many matches in the stand-alone table.

Select a feature in the *Rivers* feature class table and click the Related Tables button to see matching records in the stand-alone table.

❹ In the *Vacant Parcels* table (on the right), make sure you can also see the PIN attribute (Figure 4-71).

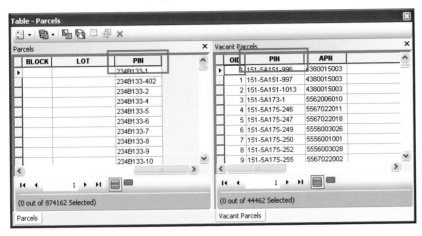

Figure 4-71

The PIN, or Parcel Identification Number, is a unique identifier. In theory, when we join the tables, each of the 44,462 records in the *Vacant Parcels* table should find a matching record in the *Parcels* table. In reality, with this many records, a perfect result is unlikely.

3) Join tables. We don't need to have the tables open to join them.

❹ Close the table window.

❸ In the Table of Contents, right-click the *Parcels* layer and choose Joins and Relates → Join.

❻ In the Join Data dialog box, make sure that the drop-down list at the top is set to "Join attributes from a table."

❹ Set drop-down list number 1 to PIN.

This specifies the key field in the *Parcels* table.

❺ Confirm that drop-down list number 2 is set to *Vacant Parcels.*

ArcMap has defaulted to this table as the join table.

❻ Make sure that drop-down list number 3 is set to PIN.

This is the key field in the *Vacant Parcels* table. ArcMap defaults to it because it's the same field name. (Key fields don't have to have the same name, however—just a compatible data type.)

❻ In the Join Options, click "Keep only matching records."

The only parcels that interest us are the vacant ones—those with matches in the *Vacant Parcels* table.

❻ Compare your dialog box to Figure 4-72.

▷ Don't click OK yet.

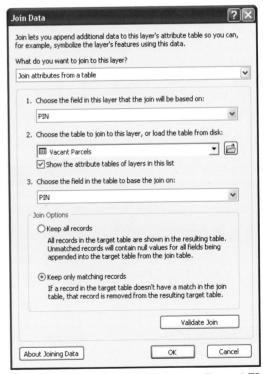

Figure 4-72

I Click Validate Join.

As the join is validated, you're prompted to create an attribute index on the PIN field in the *Vacant Parcels* table (Figure 4-73).

Figure 4-73

J On the Create Index prompt, click Yes.

It takes a while to validate the join. When the process is done, you'll get a message (Figure 4-74).

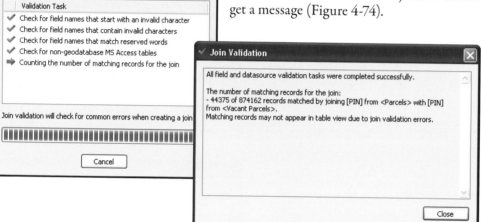

Figure 4-74

K On the Join Validation message, click Close.

L On the Join Data dialog box, click OK.

Attribute indexes

Creating an attribute index may improve the speed of the join operation—depending on the size of the table being joined—and is generally a good idea. An index also improves performance when the indexed field is queried or used to symbolize features.

Indexed fields are marked with an asterisk in attribute tables. To add or remove an attribute index from a field, open the feature class or table properties and click the Indexes tab. (The OBJECTID field in geodatabase feature classes is indexed automatically.)

When the operation is done, the *Parcels* layer redraws in the map. Because you chose to keep only the matching records, vacant parcels are the only ones that draw (Figure 4-75).

ⓜ Open the attribute table for the *Parcels* layer.

ⓝ At the bottom of the table, click the Move to end of table button ▸⁌.

ⓞ Scroll across the attributes.

The joined table has the attributes of both tables. It has 44,375 records, which means that 87 records from the *Vacant Parcels* table were unmatched (Figure 4-76). That's a match rate of 99.8 percent.

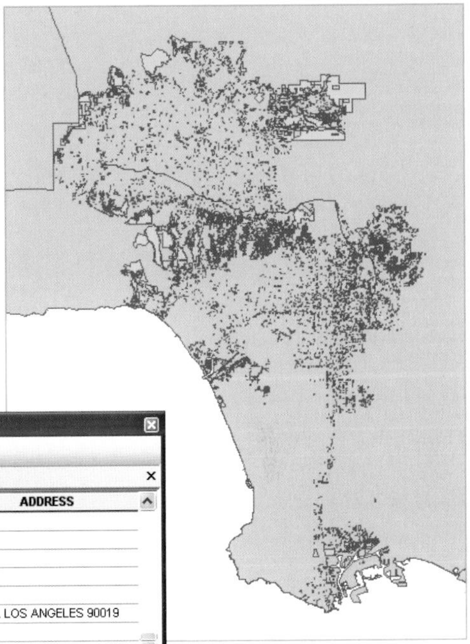

Figure 4-75

The asterisk on the PIN field means that it's indexed.

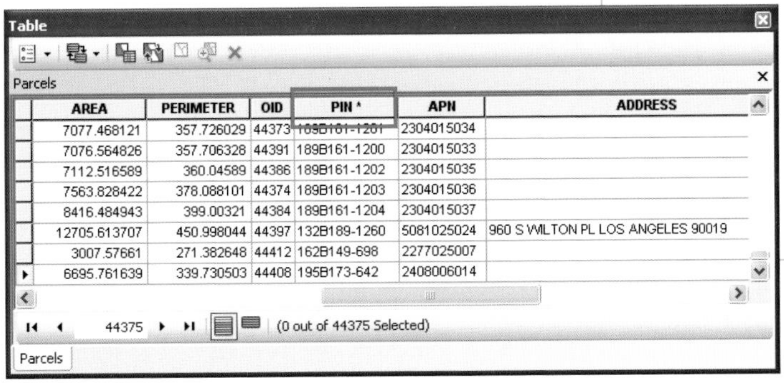

Table

Parcels

	AREA	PERIMETER	OID	PIN *	APN	ADDRESS
	7077.468121	357.726029	44373	189B161-1201	2304015034	
	7076.564826	357.706328	44391	189B161-1200	2304015033	
	7112.516589	360.04589	44386	189B161-1202	2304015035	
	7563.828422	378.088101	44374	189B161-1203	2304015036	
	8416.484943	399.00321	44384	189B161-1204	2304015037	
	12705.613707	450.998044	44397	132B189-1260	5081025024	960 S WILTON PL LOS ANGELES 90019
	3007.57661	271.382648	44412	162B149-698	2277025007	
	6695.761639	339.730503	44408	195B173-642	2408006014	

⏮ ◀ 44375 ▶ ▶⏭ ▦ ▭ (0 out of 44375 Selected)

Parcels

Figure 4-76

ⓟ Close the table.

4) Copy features. We'll create a new feature class in ReadyData, copying just the vacant parcel features from the joined table.

ⓐ Open the Copy Features tool.

▷ How? Open the Search window, then click the Copy Features tool. (If the tool isn't in the list, type its name in the search box.)

ⓑ In the Copy Features dialog box, set the Input Features to the *Parcels* layer.

ⓒ In the Output Feature Class parameter, clear the path, type **VacantParcels**, and press Tab.

ⓓ Compare your dialog box to Figure 4-77, then click OK.

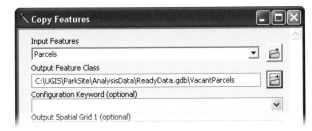

Figure 4-77

When the tool is finished, the new feature class appears in ReadyData and a layer is added to the map.

E In the Table of Contents, remove the (old) *Parcels* layer and the (old) *Vacant Parcels* table.

F At the top of the Table of Contents, click the List By Drawing Order button.

G Open the attribute table for the (new) *VacantParcels* layer and scroll across the attributes.

The virtual attributes of the joined table are now permanent attributes of the new feature class. The field names are prefixed with the name of their original source table. (If we needed to work with these attributes, we would alias the field names, but it's not an issue for us.)

H Close the table.

5) Dissolve vacant parcels. We still need to calculate acreage for the vacant parcels (Row 3 of the requirements table). But at this point we might have a new thought: what if there are some adjacent vacant parcels that are smaller than an acre individually but would be more than an acre if combined? We might be missing out on some potential park sites simply because of property lines (Figure 4-78).

The example shows three adjacent vacant parcels. Separately, each is less than an acre, but together they're more than an acre. We want to combine them into a potential park site.

Figure 4-78

To address this possibility, we'll dissolve the *VacantParcels* layer, which will create single features out of parcels sharing a common boundary. In the process, we're going to lose attributes. We don't really need the attributes though—at least, we don't need them as much as we need good park locations.

A From the ArcMap main menu, choose Geoprocessing → Dissolve.

B In the Dissolve dialog box, set the Input Features to the *VacantParcels* layer.

C Change the output feature class name to **VacantParcels2**.

D In the Dissolve Fields area, leave all the fields unchecked.

We're dissolving purely on the parcel geometry. (See *Dissolving features* on page 140.)

ⓔ At the bottom of the dialog box, uncheck the "Create multipart features" box.

If we allowed multipart features, the output feature class would consist of a single multipart polygon. We'd have one great big discontinuous vacant parcel.

ⓕ Compare your dialog box to Figure 4-79, then click OK.

When the dissolve is complete, *VacantParcels2* is added to ReadyData and a layer is added to the map.

ⓖ In the Table of Contents, turn *VacantParcels2* off and on a few times.

Even with the map zoomed out to the city limits, you can see some areas where parcel boundaries have disappeared.

ⓗ In the Table of Contents, remove the (old) *VacantParcels* layer.

ⓘ Open the attribute table for the (new) *VacantParcels2* layer.

It has 15,802 records: about a third as many as the undissolved layer. The only attributes are the four managed by ArcMap (Figure 4-80).

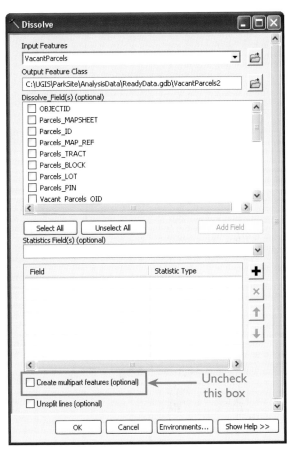

Figure 4-79

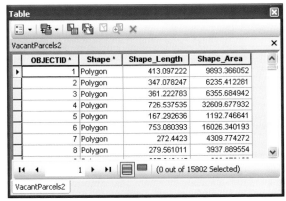

Figure 4-80

6) **Calculate acreage.** Let's get back to acreage. ArcMap has recalculated the Shape_Area field for all dissolved parcels, but the units are in square feet. We have to make a calculation to convert them to acres.

ⓐ Add a new field to the *VacantParcels2* table.

▷ How? Click the Table Options button at the top of the table window and choose Add Field.

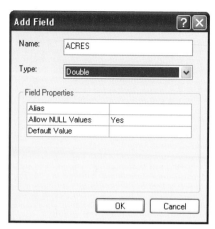

Figure 4-81

B Give the new field these properties::
- Name: **ACRES**
- Type: Double

C Compare your Add Field dialog box to Figure 4-81, then click OK.

D In the table, right-click the ACRES field heading and choose Calculate Geometry.

E In the Calculate Geometry dialog box, make sure the Property is Area and the coordinate system option is "Use coordinate system of the data source."

F Click the Units drop-down arrow and choose Acres US [ac].

G Compare your Calculate Geometry dialog box to Figure 4-82, then click OK.

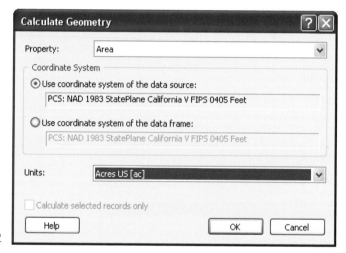

Figure 4-82

ArcMap calculates the acreage for each record and adds the values to the table (Figure 4-83).

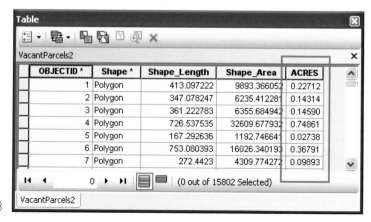

Figure 4-83

Ⓗ Close the table.

7) Clean up the ReadyData geodatabase. We'll get rid of the undissolved *VacantParcels* feature class, and rename the dissolved one. Then we'll edit the item description for this feature class.

Ⓐ In the Catalog window, expand ReadyData.gdb.

Ⓑ Right-click *VacantParcels* and choose Delete.

  ▷ Make sure it's *VacantParcels,* not *VacantParcels2*!

Ⓒ Click Yes on the prompt to confirm your deletion.

The feature class is deleted from the geodatabase.

Ⓓ In the Catalog window, right-click *VacantParcels2* and choose Rename.

Ⓔ Rename the feature class **VacantParcels** and press Enter.

Ⓕ Right-click the renamed *VacantParcels* feature class and choose Item Description.

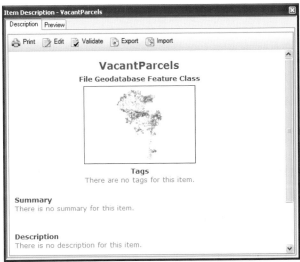

Figure 4-84

Except for the feature class name and file format, the Item Description is empty.

Ⓖ In the Item Description window, click the Preview tab.

Ⓗ In the row of tools at the top of the window, click the Create Thumbnail button 📇.

Ⓘ Click the Description tab.

The Item Description is updated with a thumbnail graphic (Figure 4-84).

Ⓙ On the Description tab, click Edit.

Ⓚ Click in the Tags box and type the following: **parcels, vacant parcels, Los Angeles** (Figure 4-85).

  ▷ Include the commas: they separate tags.

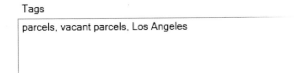

Figure 4-85

Ⓛ Scroll down, if necessary. Click in the summary box and type: **This dataset of land parcels in the city of Los Angeles supports queries on the size of contiguous vacant parcels.**

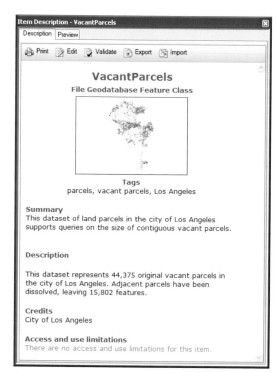

Figure 4-86

Figure 4-87

M Click in the Description box and type: `This dataset represents 44,375 original vacant parcels in the city of Los Angeles. Adjacent parcels have been dissolved, leaving 15,802 features.`

N Click in the Credits box and type: `City of Los Angeles`.

O At the top of the Description tab, click Save.

The Item Description should look like Figure 4-86.

P Close the Item Description.

Q Optionally, in the Table of Contents, rename the *VacantParcels2* layer to `VacantParcels`.

R In the Catalog window, confirm that your ReadyData geodatabase looks like Figure 4-87.

S Save the map document.

T Close the data requirements table, if you have it open.

U Exit ArcMap.

The project geodatabase has been created and the data has been prepared for the analysis. In Lesson 3, while looking at park data, we saw that the shape of Pecan Playground wasn't very accurate and that Vista Hermosa Park was a missing feature. In the next lesson, we'll make edits to park features and attributes.

Deleting items from the catalog

Items you delete from ArcCatalog, or from the Catalog window in ArcMap, are permanently removed from your computer. They cannot be retrieved from the Recycle Bin in Windows or "undeleted" by any ArcGIS command. This is true for any item shown in the catalog: datasets, toolboxes, map documents, and files of all other kinds. Be very careful about deletions and always keep backup copies of your work. In the Catalog window, you can easily copy an item (including a geodatabase or a folder) by right-clicking it and choosing Copy.

Lesson

5 Edit data

DATA INTEGRITY IS CRUCIAL IN GIS, so it shouldn't be easy to change feature shapes or locations by accident. In ArcMap, before you can make a spatial change to a feature, you have to start an edit session. That's the first safeguard the software applies to make sure you really mean to do what you're doing. Within an edit session, if you make a mistake, you can recover with the Undo button, which lets you undo edits one by one in reverse order. If the mistakes are too complicated to resolve, you can stop an edit session without saving edits.

Ideally, data would come to us free of errors or inconsistencies, but that seldom happens. Careful exploration of data usually reveals imperfections, which arise for all kinds of reasons. Sometimes data is captured or created incorrectly to begin with. Sometimes errors creep in with subsequent data processing. Sometimes the data is perfectly good until the world changes. And sometimes the data isn't wrong at all: it's just generalized for use at a certain scale and inappropriate at other scales. For example, the parks data we're using was really designed for medium-scale maps, more or less in the range of 1:100,000 to 1:250,000. When the features are examined at much larger scales, it's not surprising that they don't conform to high-resolution imagery.

ArcMap has many tools for editing features in specialized ways, as well as methods for evaluating and maintaining data integrity. Data editing, however, isn't the focus of our project. We're correcting one park feature and creating another because, during our data exploration, we saw a need to do so. Then we'll move ahead with the analysis.

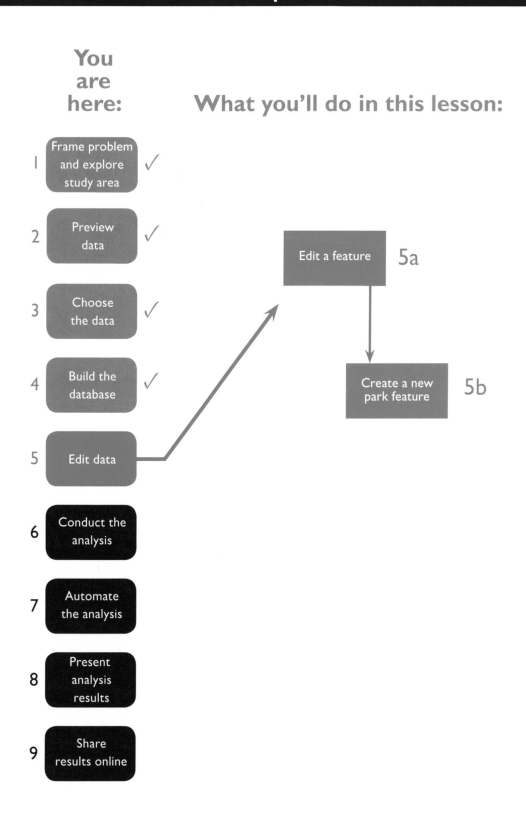

You are here:

What you'll do in this lesson:

1 Frame problem and explore study area ✓

2 Preview data ✓

3 Choose the data ✓

4 Build the database ✓

5 Edit data

Edit a feature 5a

Create a new park feature 5b

6 Conduct the analysis

7 Automate the analysis

8 Present analysis results

9 Share results online

Exercise 5a: Edit a feature

In Lesson 3, we saw that Pecan Playground, a park near Dodger Stadium, didn't match up very well against imagery. In this exercise, we'll edit the park boundary and update its acreage attribute.

1) **Add data to ArcMap.** We need the parks data and a basemap of imagery.

Ⓐ Start ArcMap and open a blank map.

Ⓑ On the Standard toolbar, click the Add Data drop-down arrow and choose Add Basemap.

Ⓒ In the Add Basemap dialog box, click on the Imagery basemap, then click Add.

Ⓓ In the Catalog window, under Folder Connections, navigate to C:\UGIS\ParkSite\AnalysisData\ReadyData.gdb.

▷ If your ReadyData geodatabase isn't up-to-date, download the results for Lesson 4 from the Understanding GIS Resource Center.

Ⓔ From ReadyData, drag and drop *Parks* onto the map.

▷ If you get a Geographic Coordinate Systems Warning, click Close.

Ⓕ Right-click the *Parks* layer and choose Zoom To Layer.

Ⓖ Optionally, change the symbol for the *Parks* layer to a shade of green (Figure 5-1).

Figure 5-1

In a little while, we'll use My Places to go to Pecan Playground. If you removed the Data Frame Tools toolbar, you need to add it.

Ⓗ If necessary, from the main menu, choose Customize → Toolbars → Data Frame Tools.

2) **Start an edit session.** Most basic editing operations are found on the Editor toolbar.

Ⓐ On the Standard toolbar, click the Editor Toolbar button ![Editor Toolbar button] to open the Editor toolbar (Figure 5-2).

You can also open the Editor toolbar in the usual way from the Customize menu.

Figure 5-2

Ⓑ Place the toolbar wherever you want.

The tools are disabled because you haven't started an edit session.

Ⓒ On the Editor toolbar, click the Editor drop-down menu and choose Start Editing.

Immediately, you encounter a warning (Figure 5-3).

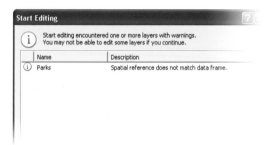

Figure 5-3

The warning tells you that the spatial reference (the coordinate system) of the *Parks* layer doesn't match the data frame. Recall from Lesson 3 (page 105) that the first layer added to the data frame sets the data frame's coordinate system. We added the *Basemap* layer first. Because its coordinate system is WGS 1984 Web Mercator, this is also the coordinate system of the data frame. The *Parks* layer has been projected to Web Mercator on the fly, but its native coordinate system is California State Plane Zone 5. That explains the mismatch.

Is this a problem? Generally speaking, operations performed on data which is projected on the fly—whether spatial query, analysis, editing, or anything else—are safe. Because data editing is a highly sensitive operation, however, the best practice is to edit data in its native coordinate system. Although this warning message wouldn't stop us from editing, we'll take this opportunity to change the data frame's coordinate system to match the *Parks* layer.

Ⓓ On the Start Editing warning, click Stop Editing.

3) **Change the data frame's coordinate system.** This operation should be familar to you from Lesson 3.

Ⓐ Open the data frame properties.
   ▷ How? At the top of the Table of Contents, double-click the data frame name (Layers).

Ⓑ In the Data Frame Properties dialog box, click the Coordinate System tab.

Ⓒ In the Select a coordinate system box, expand the Layers folder, and expand the Parks folder.

Ⓓ Click the coordinate system name inside this folder.

The current coordinate system of the data frame changes to California State Plane Zone 5. This is the native coordinate system of the *Parks* layer, which is what we want (Figure 5-4).

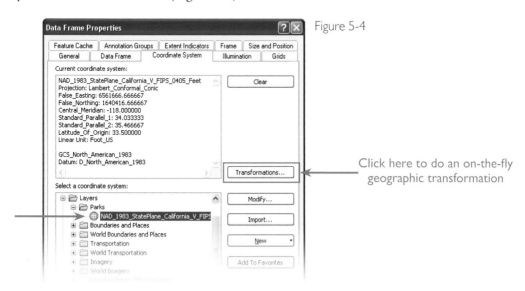

Figure 5-4

Click here to do an on-the-fly geographic transformation

We now have a mismatch between the geographic coordinate system of the data frame (NAD 1983) and the sublayers of the *Basemap* layer (WGS 1984). We can fix this with a geographic transformation, like the one we did to the *Mjr_rd* dataset in Lesson 4. This time, we'll do the transformation on the fly in the data frame.

On the right side of the dialog box, click Transformations.

Ⓔ In the Geographic Coordinate System Transformations dialog box, in the Convert from box, click GCS_WGS_1984.

Ⓕ Confirm that the Into box is set to GCS_North_American_1983.

Ⓖ In the Using box, click the drop-down arrow and choose NAD_1983_To_WGS_1984_5.

Ⓗ Confirm that your settings match Figure 5-5, then click OK.

Ⓘ Click OK on the Data Frame Properties dialog box.

You don't get a geographic coordinate systems warning because everything is reconciled within the data frame.

Figure 5-5

4) Zoom to Pecan Playground. Let's zoom in to the feature we want to edit.

Ⓐ Optionally, maximize the ArcMap window.

Ⓑ On the Data Frame Tools toolbar, click the My Places button 🏳 (Figure 5-6).

Figure 5-6

Ⓒ In the My Places dialog box, right-click Pecan Playground and choose Zoom To.

▷ If you don't have this place, download the Lesson 4 results from the Understanding GIS Resource Center. Then, in the My Places dialog box, click the Load button and open the file *Places.dat* from the MapsAndMore folder.

Ⓓ Close My Places.

We should change the park symbol so we can see what we're doing while we're editing.

Figure 5-7

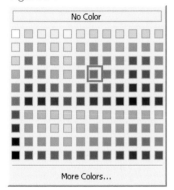

Ⓔ Open the Symbol Selector for the *Parks* layer and change the fill color to No Color.

Ⓕ Change the outline color to Quetzel Green (Figure 5-7) and the outline width to 2. Click OK on the Symbol Selector.

5) Edit the Pecan Playground feature. We're ready to start editing. Don't worry too much about your results—it should be fairly easy to make the feature better than it is now, and that's good enough.

Ⓐ On the Editor toolbar, click the Editor drop-down menu and choose Start Editing.

Several tools on the Editor toolbar become enabled and a Create Features window opens (Figure 5-8). We don't need this window now because we're editing an existing feature, not creating any new ones. We'll look at it again in the next exercise.

Ⓑ Close the Create Features window.

Figure 5-8

In the Create Features window, you set some editing behavior and default properties for new features.

C Zoom and pan as needed so you see the whole block, as in Figure 5-9.

The feature as drawn doesn't include the swimming pool or the play area at the north end of the block. Also, its boundaries extend into the street. (The southwest corner of the block is correctly excluded: this is a school.) After editing, Pecan Playground should look like Figure 5-10.

Swimming
pool   Playground

Basemap layers are updated
periodically, so your imagery
may look different.

Figure 5-9

Figure 5-10

D On the Editor toolbar, click the Edit tool ▶.

E Click anywhere inside the Pecan Playground feature to select it.

The feature outline turns blue.

F On the Editor toolbar, click the Edit Vertices button ▧.

Two things happen. One is that a new toolbar, the Edit Vertices toolbar, is added (Figure 5-11).

Figure 5-11

The other is that the feature is now marked with a number of small green squares (and one red one). This representation of the feature is called its "edit sketch." The squares, or vertices, are coordinate pairs that define the feature's shape and position. The red vertex is the last one added when the feature was created or edited. In Figure 5-12, though not on your screen, the vertices are numbered for reference. For more information, see the topic *Edit sketches and snapping* on page 181.

Figure 5-12

As you edit the feature, you may find it useful to zoom and pan with shortcut keys.

Important editing shortcut keys

To navigate while editing vertices, use these shortcut keys:

    Z to zoom in; X to zoom out; C to pan

    B for zoom/pan (click and drag to zoom; right-click and drag to pan)

Clicking navigation tools (or other tools) on the Standard toolbar interrupts vertex editing. If you accidentally interrupt vertex editing, click the Edit Vertices button on the Editor toolbar to resume it.

G On the Edit Vertices toolbar, make sure the Modify Sketch Vertices tool ▷ is selected.

H Place the mouse pointer over Vertex 2, which is in a street intersection.

When the mouse pointer is over the vertex, it changes to a four-headed arrow.

I Click and drag the vertex to where it should go—the southeastern corner of the soccer field—then release the mouse button (Figure 5-13).

▹ If you make a mistake, you can recover with the Undo button ↶ on the Standard toolbar.

Figure 5-13

Put the mouse pointer over the vertex you want to edit

Click and drag to the desired location

Release the mouse button

The blue highlight continues to show the original shape of the feature.

J Place the mouse pointer over Vertex 0 and move the vertex to the northeastern corner of the park (Figure 5-14).

The eastern boundary now has a kink because of Vertex 1. This vertex is superfluous.

K On the Edit Vertices toolbar, click the Delete Vertex tool ▷.

L Drag a small box around Vertex 1 (Figure 5-15), then release the mouse button to delete the vertex.

Figure 5-14

Figure 5-15

The eastern side of the park should now be straight, as in Figure 5-16.

Figure 5-16

Figure 5-17

Figure 5-18

M On the Edit Vertices toolbar, click the Modify Sketch Vertices tool ▷ and move Vertex 7 (the red vertex) to the northwest corner of the park (Figure 5-17).

N Move Vertex 5 to where the baseball diamond meets the school-yard (Figure 5-18).

O Click the Delete Vertex tool ▷— and delete Vertex 6 to straighten the western side of the park (Figure 5-19).

P Click the Modify Sketch Vertices tool ▷. Move Vertex 3 to the southwestern corner of the soccer field (Figure 5-20).

Figure 5-18

Figure 5-19

Figure 5-20

6) **Explore snapping.** When you move a vertex near another feature, the vertex connects to it automatically. This behavior, called snapping, helps prevent errors like small gaps and overshoots. A vertex will snap to features in its own layer, including its own edit sketch, as well as to features in other layers. You may have seen this behavior already with one or two of the vertices you've moved.

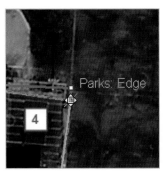

Figure 5-21

Ⓐ Place the mouse pointer over Vertex 4, the only one you haven't edited yet. (It's at the northeastern corner of the school.)

This vertex is already at a good location. We're just going to explore snapping behavior a little bit.

Ⓑ Without releasing the mouse button, drag the vertex a small distance in various directions.

As you move south or west along the edges of the feature sketch, you'll see the snap tip "Parks: Edge" appear next to the mouse pointer (Figure 5-21). The snap tip tells you that you're within the range, or snapping tolerance, of an edge in the *Parks* layer. As you move back toward your original position, you'll see the snap tip "Parks: Vertex." Moving in any direction away from the feature, you'll see no snap tip.

Ⓒ Drag the vertex back towards its original position. When you see the "Parks: Vertex" snap tip, release the mouse button.

Ⓓ Hold down the Z key on the keyboard and zoom in on Vertex 4.

▷ If you zoom in too far and the imagery disappears, hold down the X key and zoom out a little bit.

Figure 5-22

Ⓔ Place the mouse pointer over Vertex 4 again. Without releasing the mouse button, drag the vertex a small distance in different directions.

The snapping behavior is the same as before, but the snapping tolerance—which is ten screen pixels by default—corresponds to a much smaller ground distance. That gives you more control over the exact placement of the vertex.

Ⓕ Again, drag the vertex back towards its original position. When you see the "Parks: Vertex" snap tip, release the mouse button.

Ⓖ Zoom out to see the entire edit sketch (Figure 5-22).

▷ You can use the X key on your keyboard, the Go Back To Previous Extent button ⬅, or My Places.

7) **Save edits.** We'll finish the sketch and save the edits.

Ⓐ On the Edit Vertices toolbar, click the Finish Sketch button.

Figure 5-23

The edited park is now a selected feature (Figure 5-23). We could still undo these edits, or stop the edit session without saving them, but we should be pretty satisfied with the new park boundary.

ⓑ On the Editor toolbar, click the Editor drop-down arrow and choose Save Edits.

8) Update attribute values. Changing the park's shape has changed its area. That means the ACRES attribute value is no longer correct.

ⓐ Open the *Parks* attribute table.

ⓑ At the bottom of the table, click the Show selected records button 🗐. Scroll across to see the geometry attributes.

ArcMap has automatically updated the SHAPE_LENGTH and SHAPE_AREA fields, but it's our job to maintain the ACRES attribute. The value stored now is just over four acres. This is the acreage of the old, pre-edited feature.

ⓒ Right-click the ACRES field heading and choose Calculate Geometry.

ⓓ Make sure the Property is set to Area and the Coordinate System is set to the data source.

ⓔ Click the Units drop-down arrow and choose Acres US [ac].

ⓕ Compare your dialog box to Figure 5-24, then click OK.

The new value is written to the table. It shouldn't be much different from the old one. Although the park was enlarged on its northern boundary, it was narrowed along three sides.

ⓖ On the Editor toolbar, click the Editor drop-down arrow and choose Save Edits.

ⓗ At the bottom of the table, click the Show all records button 🗐.

ⓘ At the top of the table, click the Clear Selection button 🗐.

ⓙ Close the attribute table.

ⓚ On the Editor toolbar, click the Editor drop-down arrow and choose Stop Editing.

Figure 5-24

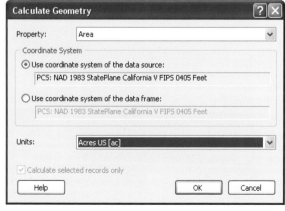

9) **Save the map document.** Your edits have been saved to the *Parks* feature class and are preserved whether or not you save the map document. We'll save the map document because we'll keep using it in Exercise 5b and you may not be continuing immediately.

ⓐ Save the map document as **Lesson5** in your MapsAndMore folder.

ⓑ If you are continuing to the next exercise now, leave ArcMap open. Otherwise, exit ArcMap.

Edit sketches and snapping

When you edit a feature, you work with its "edit sketch." An edit sketch is a representation of the feature with its coordinate anatomy exposed. It is made up of vertices, edges, and ends. Vertices define the shape and location of a feature, and usually mark a change in angle of the feature's shape. Edges, also called segments, are lines that connect vertices. Ends are the vertices at either end of a line feature. The special vocabulary makes clear that vertices, edges, and ends are not features in and of themselves, but rather components of features.

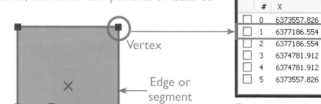

Each vertex is an x,y coordinate pair.

A park feature   Its edit sketch

Every vertex is an x,y coordinate pair. You can view (and edit) the list of coordinate pairs by clicking the Sketch Properties button 🔼 on the Editor Toolbar.

Vertices and edges can be moved with the mouse, or moved and edited in other ways with context menu commands.

As you move a vertex or edge, it will snap (connect automatically) to parts of nearby features. Snapping behavior is controlled with the Snapping toolbar. By default, a vertex or edge that you're moving will snap to edges, vertices, ends, or point features in any feature-based layer in the data frame. To turn snapping off and on for any of these elements, click its button on the toolbar.

Moving an edge

Moving a vertex

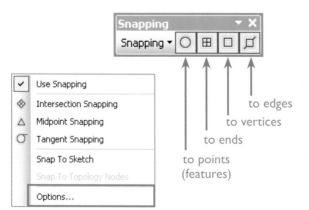

The Snapping drop-down menu on the toolbar lets you turn all snapping off. It also lets you control snapping to some additional elements and to the edit sketch of a new feature as you draw it.

The Options menu choice opens a dialog box that lets you change the snapping tolerance from the default value of ten pixels. (The snapping tolerance is the distance within which snapping takes effect.) You can also turn snap tips on or off and set the way snap tips are formatted.

Exercise 5b: Create a new park feature

Another local park, Vista Hermosa Park, was represented by a raster dataset in our source data, but not by a feature in any of the parks feature classes. In this exercise, we'll create it by tracing its outline against the basemap image. The process is very similar to the editing that you did in the last exercise.

After creating the feature, we'll add attribute values for it. We'll also edit the attribute values for the two park features we loaded in Lesson 4.

1) **Start an edit session.** We'll zoom to the location of the missing park and start an edit session.

Ⓐ If necessary, start ArcMap and open Lesson5 from the list of recent maps.

Ⓑ Use My Places to zoom to Vista Hermosa Park, then close My Places.

The park itself doesn't take up the entire L-shaped lot, just the portion shown by the dotted outline in Figure 5-25.

Figure 5-25

You won't see the dotted line on your map—it was added here for reference.

Ⓒ Zoom in to an area more or less matching Figure 5-25.

Ⓓ In the Table of Contents, right-click the *Parks* layer and choose
Edit Features → Start Editing.

This is another way to start an edit session, and it opens the Create
Features window again.

Ⓔ In the Create Features window, select the Parks symbol (Figure
5-26).

By clicking the symbol, you've chosen a "template." The template
determines the properties that new features you create will have.
Because our map only has one editable layer, Parks is the only
template available.

Ⓕ Right-click the Parks template and choose Properties.

The Template Properties dialog box lets you view and change the
properties of the template (Figure 5-27).

Figure 5-26

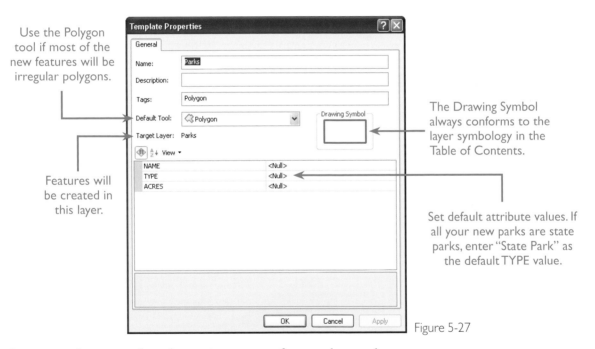

Use the Polygon
tool if most of the
new features will be
irregular polygons.

Features will
be created in
this layer.

The Drawing Symbol
always conforms to the
layer symbology in the
Table of Contents.

Set default attribute values. If
all your new parks are state
parks, enter "State Park" as
the default TYPE value.

Figure 5-27

In our case, because we're only creating one new feature, the template
isn't that useful to us.

Ⓖ Close the Template Properties dialog box.

Ⓗ Auto hide the Create Features window.

▷ How? Click the pushpin icon in the upper right corner of the
window's title bar.

2) Create the park feature. We're ready to draw the boundary of Vista Hermosa Park. Again, our purposes don't require extremely high accuracy. We just want a boundary that approximately conforms to the yellow dotted outline in Figure 5-25. If you decide to zoom in and work at large scale, you may be able to see and follow the fence line that encloses the park. (Remember to use the editing shortcut keys.)

Ⓐ Move the mouse pointer over the map.

The cursor becomes a crosshairs.

Ⓑ Choose a beginning point. This can be anywhere, but corners are usually good—for example, you might use the southeast corner of the soccer field.

Ⓒ Click once to start drawing the feature.

A red vertex is added at the location where you clicked and a semi-transparent Feature Construction toolbar appears (Figure 5-28).

Ⓓ Move the mouse in the direction you want to go. (Don't drag, just move.)

As you move, the cursor remains connected to the vertex by a purplish line (Figure 5-29). As long as you're moving in a straight line, you don't need to add a vertex.

Figure 5-28

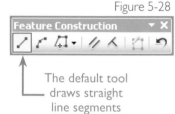

The default tool draws straight line segments

Figure 5-29

❺ When you come to a place where you need to change direction, click to add a second vertex.

The new vertex is red (it's the last one added) and the first vertex turns green.

❻ Trace the boundary of the park as well as you can, following fence lines, property lines, and sidewalks.

▷ If you add a vertex in a bad place, click the Undo Add Vertex button ↰ on the Feature Construction toolbar.

❼ When the polygon is complete, on the Feature Construction toolbar, click the Finish Sketch button ⬚.

Your new feature, still selected, should resemble Figure 5-30.

Figure 5-30

Again, the result doesn't have to be perfect. If you want to make some changes, however, click the Edit Vertices ⬚ button on the Editor toolbar and go ahead. If you're really dissatisfied with your feature, press the Delete key on the keyboard to get rid of it, then start over.

❽ Assuming you're satisfied with the new park, on the Editor toolbar, click the Editor drop-down arrow and choose Save Edits.

The Feature Construction toolbar

Tools on the Feature Construction toolbar let you control the properties and behavior of line segments. For example, you can draw true curves or constrain line segments to be parallel or perpendicular to existing segments. You can also right-click while drawing a line segment to open a context menu with more options. If the Feature Construction toolbar gets in your way as you move around the screen, you can drag it to another location without interrupting your editing.

3) Edit attribute values. We've created the Vista Hermosa Park feature, but we haven't given it attribute values.

Ⓐ On the Editor toolbar, click the Attributes button ▦.

The Attributes window opens, showing the attributes and values for this feature (Figure 5-31).

Figure 5-31

Figure 5-32

Values in software-managed fields are grayed-out and can't be edited. (Your length and area values won't match those in the figure. Your OBJECTID value may be different, too.)

Ⓑ In the Attributes window, click the <Null> value next to NAME and enter **Vista Hermosa**.

Ⓒ Click the <Null> value next to TYPE and enter **Local park or recreation area** (Figure 5-32).

Ⓓ Close the Attributes window.

Ⓔ Save your edits.

We still need to calculate the acreage for Vista Hermosa Park and for the two park features we loaded in Lesson 4 (which didn't have an acreage attribute). We could find these records by scrolling to the bottom of the *Parks* attribute table, but a more systematic way is to query for Null values in the ACRES field.

Ⓕ Open the *Parks* attribute table.

Ⓖ At the top of the table, click the Select By Attributes button ▦.

H In the Select by Attributes dialog box, make a query expression as follows:

- Make sure the Method is set to "Create a new selection."
- Double-click the "ACRES" field.
- Click the "Is" button (or type the word **IS**).
- Click Get Unique Values and double-click NULL.

I Confirm that your expression matches Figure 5-33, then click Apply.

J Close the Select by Attributes dialog box.

K In the table, click the Show selected records button ▤.

Three records should be selected (Figure 5-34).

Figure 5-33

Table

Parks

OBJECTID ^	Shape ^	NAME	TYPE	ACRES	Shape_Len
168	Polygon	Los Angeles	State Historic Park	<Null>	7426.87
169	Polygon	Rio de Los Angeles	State Recreation Area	<Null>	5304.01
170	Polygon	Vista Hermosa	Local park or recreation area	<Null>	2769.90

Figure 5-34

L In the table, calculate the geometry for the ACRES field.

▷ How? Right-click the ACRES field heading and choose Calculate Geometry. Set the units to Acres US [ac] and accept the other defaults.

The results should be about 32 acres for Los Angeles State Historic Park, 36 acres for Rio de Los Angeles State Recreation Area, and nine or so acres for Vista Hermosa Park (Figure 5-35).

NAME	TYPE	ACRES
Los Angeles	State Historic Park	32.134263
Rio de Los Angeles	State Recreation Area	36.752289
Vista Hermosa	Local park or recreation area	9.376164

Figure 5-35

The TYPE values for Los Angeles and Rio de Los Angeles (the two parks we loaded) are nonstandard in our table. All other state parks have the value "State or local park or forest."

M In the table, right-click on the gray square next to the Vista Hermosa record and choose Unselect.

That leaves two selected records.

Figure 5-36

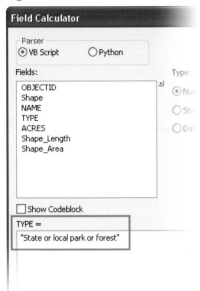

N Right-click the TYPE field heading and choose Field Calculator.

O In the expression box at the bottom of the Field Calculator, type
"State or local park or forest".

  ▷ The quotation marks are required for a string value.

P Compare your dialog box to Figure 5-36, then click OK.

The values for the two selected records are updated (Figure 5-37).

NAME	TYPE	ACRES
Los Angeles	State or local park or forest	32.134263
Rio de Los Angeles	State or local park or forest	36.752289

Figure 5-37

4) **Stop editing.** We're finished with edits to the *Parks* layer.

A Save your edits.

B In the table, clear the selection 🗹, then show all records 📑.

C Stop editing and close the attribute table.

Editing layers in different workspaces

During an edit session, you can only edit layers that derive from a single workspace—that is, from the same geodatabase or the same folder. If your layers point to different workspaces, you'll be prompted to choose which workspace you want to make editable.

5) **Zoom out and save the map.** We'll take a quick look at our work from a smaller scale, then save the map.

A Resymbolize the *Parks* layer to give it a solid green fill color and a thinner outline, like **0.5**. Change the outline color, too, if you want.

B Open My Places.

C In the My Places dialog box, hold down the Shift key and select both Vista Hermosa Park and Pecan Playground (Figure 5-38).

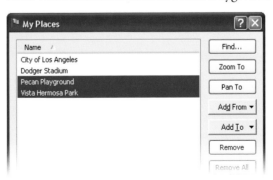

Figure 5-38

D Click Zoom To, then close My Places.

On the map, you're zoomed to an extent that includes both parks (Figure 5-39).

Vista Hermosa Park ⟶

⟵ Pecan Playground

Figure 5-39

E Remove the Editor toolbar, the Data Frame Tools toolbar, and any other specialized toolbars that may be open.

As mentioned earlier, edits are written to the source data at the time you save them. Although we're making a practice of saving map documents, your park edits are already saved.

F Save the map, then exit ArcMap.

All our data preparations are complete. In the next lesson, we'll look for suitable park sites.

Lesson

6 Conduct the analysis

NOW WE COME TO THE POINT

of our preparatory work: the analysis. Our original feature class of parcels was like an assembled 874,000-piece jigsaw puzzle covering the entire city, each piece representing an individually owned property. In Lesson 4, we cut this down to about 45,000 parcels (those that were vacant), then to 15,000 (those that were adjacent). Now we'll continue taking pieces away, removing non-suitable parcels until we're left with just those candidates that meet the criteria established in Lesson 2:

- On a vacant land parcel one or more acres in size
- Within the LA city limits
- Within 0.75 miles of the LA River (the closer the better)
- At least 0.25 miles from the nearest park
- In a neighborhood where population density is at least 8,000 people per square mile
- Where at least 25 percent of the population is under 18
- Where median household income is $50,000 or less
- Preferably serving the most people within a quarter-mile radius

But how does this translate into a method and a workflow? How do we take these requirements and match them to appropriate geoprocessing tools in the right sequence? Planning a workflow means identifying the tasks, finding a suitable tool for each, and understanding the inputs and outputs so we can order the operations in a logical, efficient way.

Most of the tasks in our analysis—and this is typical—fall into just a few categories: distance problems (what's near what), topological problems (what crosses, touches, or contains what), attribute query problems (what has this or that value), and overlay problems (what areas are common to features in different layers). Each of these problem categories is associated with its own special set of tools.

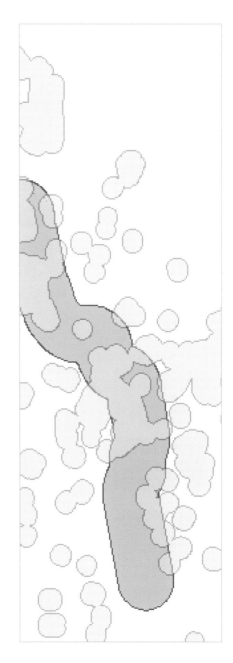

Lesson Six roadmap

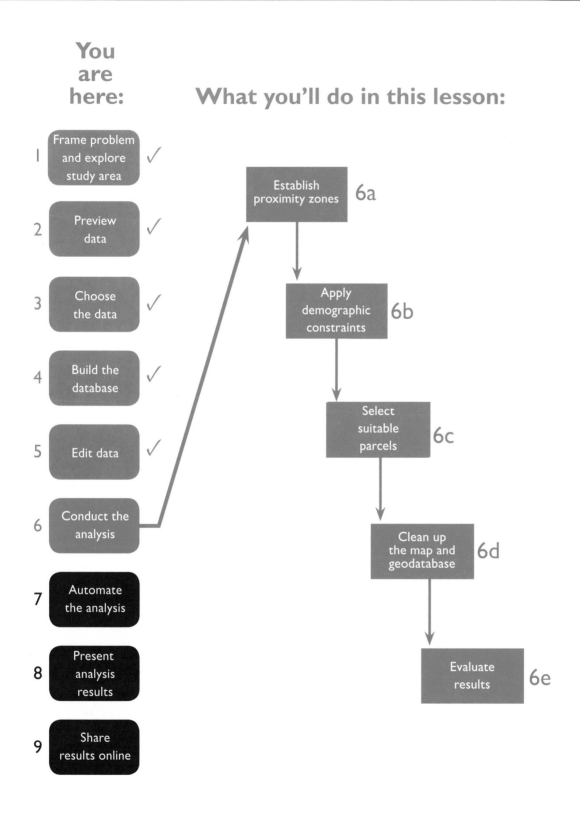

You are here:

1 Frame problem and explore study area ✓

2 Preview data ✓

3 Choose the data ✓

4 Build the database ✓

5 Edit data ✓

6 Conduct the analysis

7 Automate the analysis

8 Present analysis results

9 Share results online

What you'll do in this lesson:

Establish proximity zones — 6a

Apply demographic constraints — 6b

Select suitable parcels — 6c

Clean up the map and geodatabase — 6d

Evaluate results — 6e

Before beginning an involved analysis, you may find it helpful to draw a rough plan or diagram. Any medium will do—paper, a software application, a whiteboard—as long as you're prepared to make revisions as you go. A typical working plan (the one we'll follow) could be sketched out to look something like Figure 6-1. This sketch depicts the broad general approach, although there are a few additional twists and turns along the way.

Figure 6-1

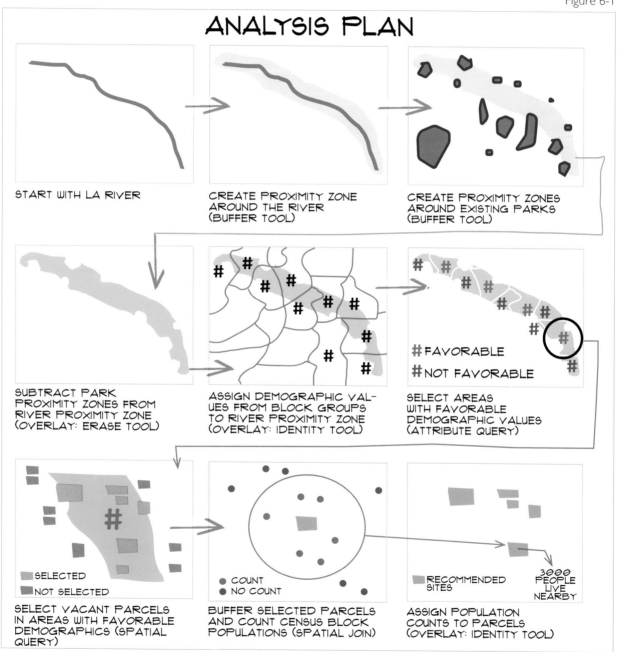

ANALYSIS PLAN

START WITH LA RIVER

CREATE PROXIMITY ZONE AROUND THE RIVER (BUFFER TOOL)

CREATE PROXIMITY ZONES AROUND EXISTING PARKS (BUFFER TOOL)

SUBTRACT PARK PROXIMITY ZONES FROM RIVER PROXIMITY ZONE (OVERLAY: ERASE TOOL)

ASSIGN DEMOGRAPHIC VAL-UES FROM BLOCK GROUPS TO RIVER PROXIMITY ZONE (OVERLAY: IDENTITY TOOL)

#FAVORABLE
#NOT FAVORABLE

SELECT AREAS WITH FAVORABLE DEMOGRAPHIC VALUES (ATTRIBUTE QUERY)

SELECTED
NOT SELECTED

SELECT VACANT PARCELS IN AREAS WITH FAVORABLE DEMOGRAPHICS (SPATIAL QUERY)

COUNT
NO COUNT

BUFFER SELECTED PARCELS AND COUNT CENSUS BLOCK POPULATIONS (SPATIAL JOIN)

RECOMMENDED SITES

3000 PEOPLE LIVE NEARBY

ASSIGN POPULATION COUNTS TO PARCELS (OVERLAY: IDENTITY TOOL)

Looking at the sketch of the analysis brings up the point that this plan represents only one of several possible approaches to the problem. The plan we're following here is, we think, a reasonable and efficient one. It uses a variety of important geoprocessing tools and has the visual benefit of letting you see the analysis unfold, as land areas are progressively stripped away from consideration. It is certainly possible, however, to reach the same or similar conclusions with different combinations of tools.

Your analysis results depend on the state of the data in the ReadyData geodatabase. If you've completed all the exercises up to now, you should be in good shape. If necessary, download the lesson results from the Understanding GIS Resource Center.

Exercise 6a: Establish proximity zones

In this exercise, we have two objectives. First, we want to define the area of interest as a ¾-mile zone around the river. Any parcels falling outside this zone will be excluded from consideration. Second, we want to draw ¼-mile zones around each park. Any parcels in the area of interest that lie within these internal exclusion zones will also be dropped. We'll proceed to use one simple operation—buffer—to significantly reduce our hunting grounds.

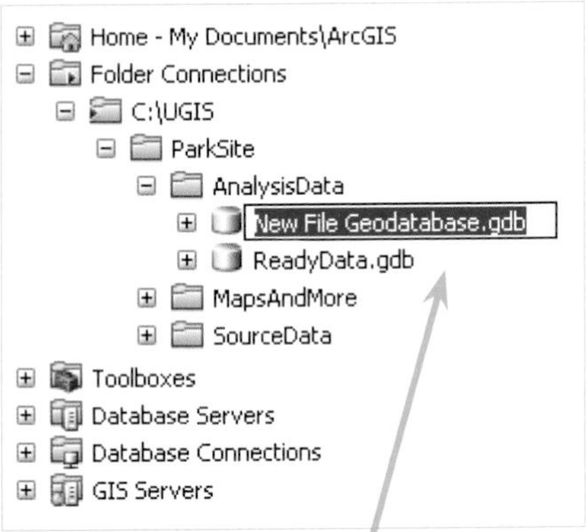

Figure 6-2

Name the new geodatabase AnalysisOutputs

1) **Create a results geodatabase.** When we generate outputs we need to store them somewhere. In Lesson 4, we decided to keep our input data in one geodatabase and our output data in another geodatabase. (See *Project database considerations* on page 122.) It's a matter of preference, but we think it's easier to keep the project organized if inputs and outputs are separated. We may want to share the input data with someone else so they can run the analysis independently. We may want to repeat the analysis with different parameters (something that will happen in Lesson 7). The more feature classes you add to a geodatabase, the harder it is to keep track of what they represent and what purpose they serve.

Ⓐ Start ArcMap and open a blank map.

Ⓑ In the Catalog window, expand Folder Connections and navigate to the folder \UGIS\ParkSite.

Ⓒ Right-click the AnalysisData folder and choose New → File Geodatabase (Figure 6-2).

Ⓓ Rename the geodatabase **AnalysisOutputs**.

▷ How? The name should be editable when you create the geodatabase. If not, you can right-click it and choose Rename.

We need to make AnalysisOutputs the default geodatabase so our geoprocessing outputs go there automatically.

🅴 In the Catalog window, right-click AnalysisOutputs and choose Make Default Geodatabase.

2) **Buffer the LA River.** A buffer is a zone around a map feature measured in distance units. We'll use the Buffer tool to create a ¾-mile proximity zone around the LA River. For more information about the Buffer tool and other analysis tools, see the topic *Essential GIS analysis* tools on pages 196–197.

🅰 In the Catalog window, under AnalysisData, expand ReadyData.gdb and add *LARiver* to the map.

Figure 6-3

Note that we added the *LARiver* layer to an empty map display without putting a contextual basemap underneath it. Also, the layer is likely symbolized in a color other than blue. During analysis, we're more interested in geoprocessing results than we are in map appearance. For that reason, we're not going to symbolize input and output layers carefully at each step along the way. We'll do it as the need arises, but often we'll just accept the ArcMap defaults.

🅱 From the main menu, choose Geoprocessing ➤ Buffer.

🅲 In the Buffer dialog box, click the Input Features drop-down arrow and choose *LARiver*.

🅳 Rename the output **LARiverBuffer**.

🅴 In the Linear Unit box, type **0.75**. Click the drop-down arrow for units and choose Miles.

🅵 Double-check your settings (Figure 6-3), then click OK.

When the tool finishes, the *LARiverBuffer* layer is added to the map.

Naming conventions revisited

In our project database considerations, we established naming conventions for the ReadyData geodatabase. We'll adhere to these conventions in AnalysisOutputs with respect to feature classes that we intend to keep and use in our final map. However, most of the geoprocessing outputs in this lesson will be "intermediate" data: essential steps in the process, but not ends in themselves. Eventually, intermediate data will be deleted from the geodatabase. With respect to naming these datasets, we'll follow an expedient approach. If the default ArcMap name is short and not confusing, we'll use it. If we need to clarify a dataset's purpose or analytical meaning in the course of operations, we'll assign it a more suitable name.

Essential GIS analysis tools

Analysis tools

The tools shown here are by no means a complete list, but they include several of the ones used most often in GIS analysis. It's hard (and probably unnecessary) to give an exact definition of what makes a tool an "analysis" tool. GIS practitioners solve problems of many different kinds, and most of these problems have facets that involve spatial relationships among geographic objects: how far is A from B, how many A's are close to B, which areas are common to A and B, how do you get from A to B, and so on. Analysis tools quantify these relationships among features and their attributes.

Query tools

Query tools answer questions of the form "Which features meet such-and-such a condition?" Select By Attributes selects features according to an attribute value or combination of values. Select By Location selects features according to their spatial relationship to features in another layer (or sometimes the same layer). Spatial relationships include intersection, containment, adjacency, and distance.

Query tools can be accessed from the Selection menu in ArcMap or as geoprocessing tools (Select Layer By Attribute and Select Layer By Location).

Proximity tools

The Buffer tool creates a feature class of polygons at a specified distance around input features. It is often used to draw an exclusion zone around features (no A's should be allowed within a mile of B), or to define an area of interest (we want to look for A's only within a mile of B).

The Create Thiessen Polygons tool creates a feature class of contiguous polygons around input point features. Each polygon's shape is defined by proximity to the nearest point. Thiessen polygons can be used to define allocation areas (which areas are closer to A than to any other point, which are closer to B than to any other point, and so on).

Select By Attributes

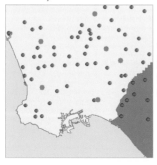

Which cities have a population of 100,000 or more?

Select By Location

Which parks are within a mile of the river?

Create Theissen Polygons

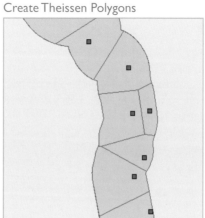

Buffer

Streams have quarter-mile buffers (buffer boundaries are dissolved).

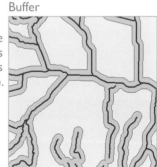

Input points are parks within a river buffer. Each output polygon encloses the area that is closer to the point it contains than to any other point.

Proximity tools (continued)

The Near tool finds the nearest feature in one or more specified layers to each feature in the input layer. It writes the distance to the nearest feature and that feature's OBJECTID as attributes to the input layer table.

Near

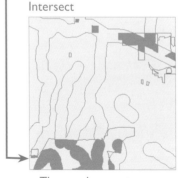

The input features are vacant parcels (symbolized in red). The near features are parks (green). New attributes specify the nearest park to each vacant parcel and its distance.

OBJECTID *	Shape *	NEAR_FID	NEAR_DIST
1	Polygon	117	1371.816526
2	Polygon	114	822.280138
3	Polygon	101	1234.614597
4	Polygon	34	2465.358086
5	Polygon	35	2012.800925

A table relate shows the attributes of the nearest park to the selected parcel.

OBJECTID *	Shape *	NAME	TYPE
117	Polygon	Greaver Oak Park	Local park or recreation area

Overlay tools

The basic purpose of overlay tools is to find common areas between different layers. Overlays answer questions about where A and B, each with unique and important attributes, overlap. In contrast to queries, which return selections of existing features, overlay tools create new features that have the attributes of both input layers.

The different overlay tools—Intersect, Union, Identity, and Erase are the most common—perform variations on the same basic process. They differ with respect to how much area from the input layers is included in the output feature class: common area only (Intersect), all area whether common or not (Union), one input layer's area only (Identity), or one input layer's area minus the common area (Erase).

Parks and stream buffers have areas of overlap.

Intersect

Spatial Join

Spatial Join is different from the overlay tools it's grouped with. It creates a new feature class, but doesn't create any features that don't already exist. It works much like a Select By Location query with an added step of joining appropriate source layer attributes to target layer features.

Spatial Join

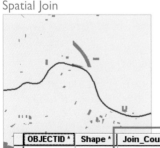

Vacant parcels (orange) within a quarter mile of the river are spatially joined to the river. In the output feature class, the river has the same geometry but new attributes: 400 vacant parcels totaling 1,232 acres satisfy the spatial join condition.

The overlay creates new features (purple) from areas of overlap. The output can be used to find park land as a percentage of riparian land or vice versa; it might also be used to analyze animal habitat or other ecological issues.

OBJECTID *	Shape *	Join_Count	TARGET_FID	NAME	ACRES	Shape_Length
1	Polyline	400	1	Los Angeles River	1232.55	166495.37503

G Confirm the presence of the new feature class by expanding the Catalog window and looking in AnalysisOutputs.gdb (Figure 6-4).

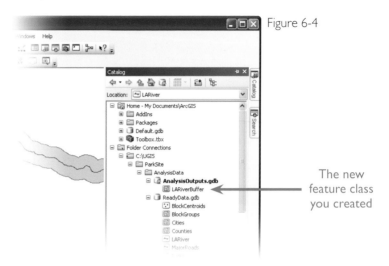

Figure 6-4

The new feature class you created

3) Buffer the parks to ¼ mile. Now we'll add the *Parks* data and draw a quarter-mile buffer around each park. These buffers will encompass areas that are close to existing parks and therefore out of consideration.

A In the Catalog window, under ReadyData, drag *Parks* to the map window.

B In the Table of Contents, right-click *LARiverBuffer* and choose Zoom To Layer.

C Turn off the *LARiver* layer.

Figure 6-5

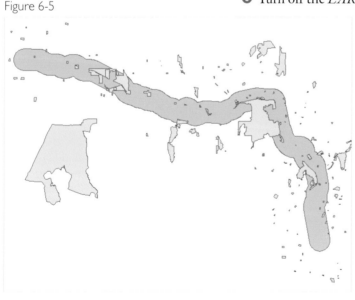

We can already see a lot of overlap between existing parks and the buffer zone (Figure 6-5). We still need to extend the exclusion zone a quarter mile further out on all sides of each park.

D From the main menu, choose Geoprocessing→Buffer.

E Set the Input Features to *Parks*.

F Accept the default Output Feature Class name of *Parks_Buffer*.

G In the Linear Unit box, type **0.25**. Set the units to Miles.

H Click the Dissolve Type drop-down arrow and choose ALL.

This setting dissolves boundaries between overlapping park buffers. The dissolution of boundaries isn't strictly required—the step that uses *Parks_Buffer* as an input would give the same result without it—but it makes it a lot easier to interpret the result, which would otherwise be a dense tangle of lines. A side effect of dissolving the buffers is that the output will consist of one multipart feature, essentially devoid of attributes. That's okay—all we need from this layer is its geometry.

Figure 6-6

○ Check your settings against Figure 6-6 and click OK.

When the tool finishes, a *Parks_Buffer* layer is added to the map and a feature class is created in AnalysisOutputs.

○ Turn off the *Parks* layer.

○ Open the layer properties for *Parks_Buffer* and click the Display tab.

○ In the Transparent % box, type **50**. Click OK.

The portion of *LARiverBuffer* not covered by the *Parks_Buffer* layer represents the areas that are still under consideration. (The layers are purple and green in Figure 6-7, but your colors may be different.)

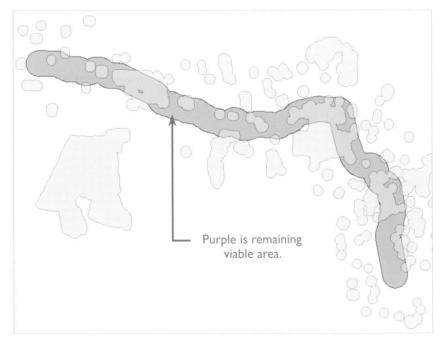

Purple is remaining viable area.

Figure 6-7

Let's think about what we have. One layer, *LARiverBuffer,* contains all the area within ¾ mile of the river. We want to exclude the *Parks_Buffer* areas from within that, but those areas are stored in a different layer. Essentially, we want to subtract one layer from another.

4) Erase park accessible areas. The tool for doing this subtraction is named Erase. This tool is a member of the family of spatial overlay operations that help make GIS such a powerful technology. Unlike the Buffer tool, the Erase tool is not available from the Geoprocessing menu, so we'll find it with a search.

A Open the Search window.
   ▷ If the Search tab isn't present, on the Standard toolbar, click the Search window button 🔍. Set the window to auto hide.
B Near the top of the Search window, click Tools to restrict your search.
C In the Search box, start to type **erase**. When "erase (analysis)" appears in the drop-down list, click on it to search for it.
D In the list of results, click Erase (Analysis), as shown in Figure 6-8, to open the tool.
E For the Input Features, choose *LARiverBuffer.*
F For the Erase Features, choose *Parks_Buffer.*
G Name the Output Feature Class **ProximityZone**.
H Check your settings against Figure 6-9, then click OK.

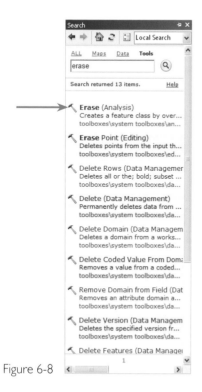

Figure 6-8

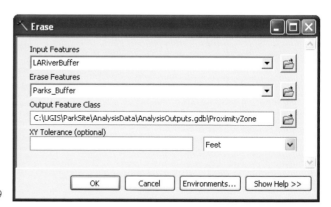

Figure 6-9

When the process is finished, the *ProximityZone* layer is added to the map.

O In the Table of Contents, turn off all layers except *ProximityZone*.

J Symbolize *ProximityZone* in a medium blue (for example, Cretean Blue). This will help it stand out against the basemap we'll add in the next step.

We're left with a reduced portion of the original buffer zone (Figure 6-10). Our search for a park is now confined to this shape.

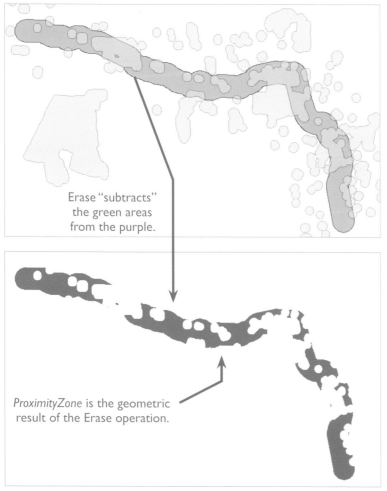

Erase "subtracts" the green areas from the purple.

ProximityZone is the geometric result of the Erase operation.

Figure 6-10

5)   Add a basemap. This new feature is quite an abstract shape, and we can't forget that it represents real territory on the ground. We'll add a basemap to get some geographic orientation.

A Click the Add Data drop-down arrow and choose Add Basemap.

B In the list of basemaps, add the Topographic basemap.

   ▷ Close the Geographic Coordinate Systems Warning if you get it.

© Open the layer properties for *ProximityZone*. Click the Display tab, if necessary.

① Change the transparency to **50**%, then click OK.

⑥ Zoom in on some of the "holes" in the *ProximityZone* layer.

You can see the effect that the existing parks (visible on the basemap) had on the creation of the feature (Figure 6-11). The map display shows us something we can understand: that our new park is going to be somewhere within the blue area.

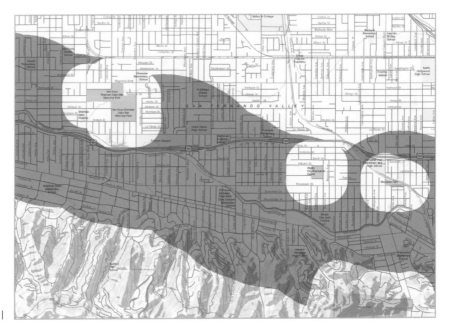

Figure 6-11

⑥ Pan along the river and look at a few more examples.

If you see some anomalies—like a hole with no park—consider that the topographic basemap may not correspond exactly to the *Parks* layer. You can turn the *Parks* layer on and off in the Table of Contents for reference.

⑥ When you're finished, remove the *Basemap* layer and zoom to the *ProximityZone* layer.

⑪ Save the map document as **Lesson6** in the MapsAndMore folder.

① If you are continuing to the next exercise now, leave ArcMap open. Otherwise, exit ArcMap.

In the next exercise we'll evaluate the neighborhood demographics within the area of interest.

Exercise 6b: Apply demographic constraints

We've isolated areas that meet two of our requirements: distance to the river and distance from parks. Still ahead are the tasks of factoring in neighborhood demographics (population density, income, and age) and evaluating parcels by size and total population served. We'll deal with the demographics in this exercise and the parcels in Exercise 6c.

Within the proximity zone, we want to find areas that have the right demographic criteria. The problem is that our blue blob, if we can call it that, doesn't include those attributes; they are found only in the *BlockGroups* feature class. What we want to do, then, is make a feature class that has the spatial area of *ProximityZone* and the attribute values of *BlockGroups*.

This again calls for an overlay. In the last exercise, we subtracted one layer from another, but overlay operations are typically more additive than subtractive. Usually, we want to find common ground between layers that have different attributes. The important attribute in *ProximityZone* is its distance to the river. The important attributes in *BlockGroups* are population density, age, and median income. By overlaying these two layers, we'll get all these important attributes in a single layer.

The overlay will cause any block group features that lie partly inside and partly outside the proximity zone to be split. What happens to the attribute values of these features? If a block group with 2,000 people is halfway in the proximity zone, should the output feature defined by the overlapping half get all 2,000 of those people, or just 1,000? By default, attribute values are copied rather than apportioned, and this can lead to anomalies with large attribute values being assigned to small slivers of geometry. We've dealt with this problem ahead of time by using block group attribute values that are statistically homogenized and assumed to be uniform across the feature. It's also possible, however, to redistribute attribute values by area during geoprocessing. See the topic *Apportioning attribute values* on page 223.

1) **Overlay block groups on the proximity zone.** We want to keep all the geometry of the *ProximityZone* layer and only as much *BlockGroups* geometry as is spatially coincident with the proximity zone. This type of overlay is called Identity: it means keep all the geometry of Layer A and just the coincident geometry of Layer B.

Ⓐ If necessary, start ArcMap and open Lesson6 from the list of recent maps.

Ⓑ In the Catalog window, add *BlockGroups* from ReadyData.gdb.

Ⓒ In the Search window, highlight and delete the existing search term. Start to type **identity**.

D Choose "identity (analysis)" from the drop-down list to locate it.

E In the list of results, click Identity (Analysis) to open the tool.

F For the Input Features, choose *ProximityZone*.

G For the Identity Features, choose *BlockGroups*.

H Accept the default Output Feature Class name of *ProximityZone_Identity*.

I Check your settings against Figure 6-12, then click OK.

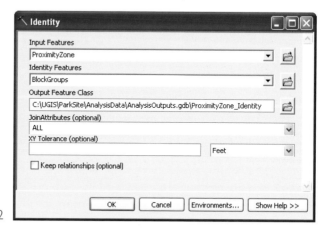

Figure 6-12

When the tool finishes, the *ProximityZone_Identity* layer is added to the map. A corresponding feature class is created in the AnalysisOutputs geodatabase.

J Zoom to the *ProximityZone_Identity* layer and turn off all other layers in the Table of Contents.

What we see in Figure 6-13 is a fractured version of the original proximity zone. *ProximityZone_Identity* has the geometry of *BlockGroups* clipped to the boundaries of *ProximityZone*. As we'll see, it has the attributes of both input layers.

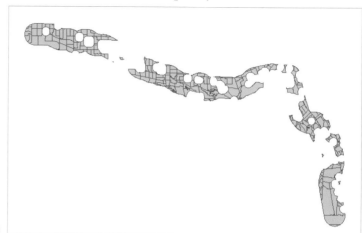

Figure 6-13

2) Examine attributes. Looking at the attributes of both our input and output layers will help us understand what we've accomplished.

Ⓐ Open the *ProximityZone* layer attribute table (Figure 6-14).

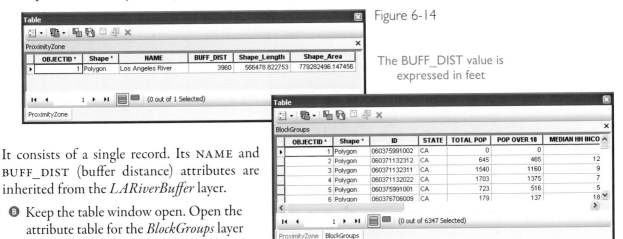

Figure 6-14

The BUFF_DIST value is expressed in feet

It consists of a single record. Its NAME and BUFF_DIST (buffer distance) attributes are inherited from the *LARiverBuffer* layer.

Ⓑ Keep the table window open. Open the attribute table for the *BlockGroups* layer (Figure 6-15). Scroll across its attributes.

This layer has 6,347 records. It has all the attributes we previewed in Lesson 2, plus those we added in Lesson 4.

Ⓒ Keep the table window open. Open the attribute table for the *ProximityZone_Identity* layer.

Ⓓ At the bottom of the table window, drag and drop the ProximityZone_Identity tab onto the blue arrow pointing upward (Figure 6-16).

Figure 6-15

Figure 6-16

These values match OBJECTID values in *BlockGroups*.

These values match OBJECTID values in *ProximityZone*.

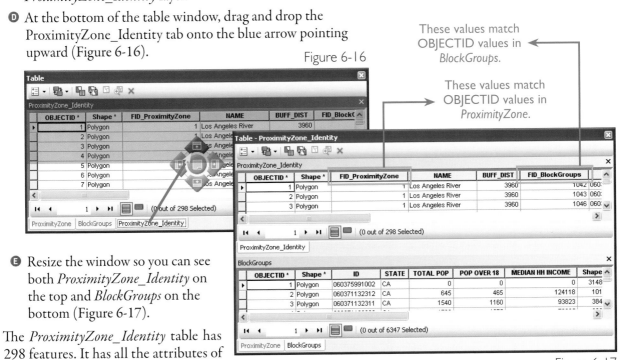

Ⓔ Resize the window so you can see both *ProximityZone_Identity* on the top and *BlockGroups* on the bottom (Figure 6-17).

The *ProximityZone_Identity* table has 298 features. It has all the attributes of both input layers, plus two ID fields that trace the features back to the input features with which they are spatially coincident.

Figure 6-17

F Keep the table window open but move it away from the map.

G Zoom in on the western end of the *ProximityZone_Identity* layer.

  ▷ A map scale of around 1:24,000 is good.

H Turn on the *BlockGroups* layer.

I On the Tools toolbar, click Select Features by Rectangle 🔲.

J Select the *ProximityZone_Identity* feature indicated in Figure 6-18, or any other feature that was cut by the proximity zone boundary.

Select this feature ——————

Figure 6-18

You can see on the map and in the tables that two features are selected: one in *ProximityZone_Identity* and one in the *BlockGroups* layer underneath.

K In the Table window, click Show Selected Records 🔲 for both tables.

L Scroll the two tables so that some common attributes line up (Figure 6-19).

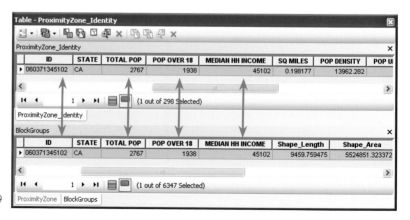

Figure 6-19

All the block group values have been copied to the spatially corresponding features in the *ProximityZone_Identity* layer. That's what we want—these cut-up block groups (some are cut up, some are intact) are the "neighborhoods" that we'll query for suitable demographics.

Note the importance of analyzing normalized attributes like density and percentage, rather than raw numbers. The POP OVER 18 value of the selected block group is 1,938. We would reasonably expect this value to be smaller in the *ProximityZone_Identity* feature, but because the attribute values are copied, it's the same. It's fair to assume, however, that the PCT UNDER 18 value would remain the same even when the feature is cut into smaller pieces.

Ⓜ On the Tools toolbar, click Clear Selected Features ⬚.

Ⓝ Close the table window.

Ⓞ Turn off the *BlockGroups* layer.

Ⓟ Zoom to the *ProximityZone_Identity* layer.

Figure 6-20

3) Select areas by demographic attributes. The *ProximityZone_Identity* layer now contains the demographic attributes of interest: population density, the percentage of children, and median household income. We want to consider areas only if they meet our thresholds for all three values:

- Population density greater than or equal to 8,000 people per square mile
- Percentage of children greater than or equal to 25%
- Median household income less than or equal to $50,000

We'll use an attribute query to select features that satisfy these conditions.

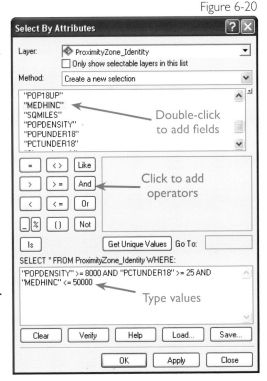

Ⓐ From the main menu, choose Selection → Select By Attributes.

Ⓑ At the top of the Select By Attributes dialog box, click the Layer drop-down menu and choose *ProximityZone_Identity*.

Ⓒ Make sure the Method is set to "Create a new selection."

Ⓓ Build the following expression:

```
"POPDENSITY" >= 8000 AND "PCTUNDER18" >= 25
            AND "MEDHINC" <= 50000
```

Combining the queries in a single statement is more efficient than making a series of selections and subselections. The AND operator ensures that features will be selected only by passing *all* the tests.

Ⓔ Check your expression against Figure 6-20, then click OK.

The map shows the features that satisfy the query. There aren't that many and most of them lie toward the ends of the proximity zone (Figure 6-21).

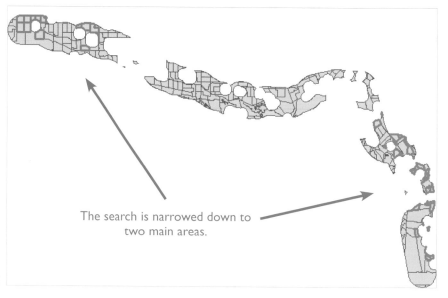

The search is narrowed down to two main areas.

Figure 6-21

F Open the *ProximityZone_Identity* attribute table.

G Confirm that there are 52 selected records, then close the table.

4) **Create a feature class from the selected features.** We'll save these selected features to a new feature class. Previously, we used the Copy Features tool for this purpose, but this time we'll use the Export Data command. The advantage of this command is its convenient location on the layer context menu. Export Data isn't incorporated into the geoprocessing framework, which means it doesn't respect environment settings, but that's not a concern in this case.

Figure 6-22

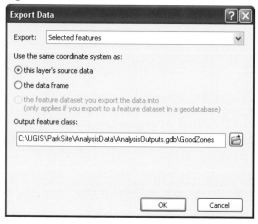

A In the Table of Contents, right-click *ProximityZone_Identity* and choose Data ➔ Export Data.

B Highlight and delete the default output feature class name (Export_Output). Replace it with **GoodZones**.

▹ Deleting the path and restoring it with the Tab key won't work in this dialog box.

C Check your settings against Figure 6-22, then click OK.

D Click Yes on the prompt to add the data as a layer.

E On the Tools toolbar, click the Clear Selected Features button ▣, then turn off the *ProximityZone_Identity* layer.

F Save the map.

G If you are continuing to the next exercise now, leave ArcMap open. Otherwise, exit ArcMap.

Exercise 6c: Select suitable parcels

We've narrowed our focus down to a layer called *GoodZones*. Having covered the demographic criteria, we'll turn our attention to the parcels. Thanks to our work in Lesson 4, we have a dataset consisting of all and only vacant parcels. From this dataset, we'll select the parcels that lie within good zones. From that selection, we'll subselect parcels that are one acre or larger: that will comprise our set of candidate park sites. Then we'll go on to do some further analysis to compare the sites by total population served and distance to the river.

1)   **Select vacant parcels within good zones.** We'll select the parcels with a spatial query.

   Ⓐ If necessary, start ArcMap and open Lesson6 from the list of recent maps.

   Ⓑ In the Catalog window, add *VacantParcels* from ReadyData.gdb.

   Ⓒ From the main menu, Choose Selection → Select by Location.

   Ⓓ At the top of the Select By Location dialog box, make sure the Selection Method is set to "select features from."

   Ⓔ Under Target Layer(s), check the box next to *VacantParcels*.

   Ⓕ Click the Source Layer drop-down arrow and choose *GoodZones*.

   Ⓖ Confirm that the spatial selection method is set to "Target layer(s) features intersect the Source layer feature."

   Ⓗ Confirm that the box to apply a search distance is unchecked.

In plain English, parcels will be selected if they lie partially or entirely within good zones.

Some parcels may cross good zone boundaries because of how the zones are clipped. Should we include parcels that straddle good zones? There's no right or wrong answer. By choosing the "intersect" spatial selection method, we're going with a generous interpretation—we'll select a parcel if any part of it lies within a good zone because we want to maximize the candidate pool. If we wanted to be restrictive (the parcel has to be *entirely* within the good zone), we'd choose a method like "Target layer(s) features are within the Source layer feature."

   Ⓘ Check your settings against **Figure 6-23**, then click OK.

Figure 6-23

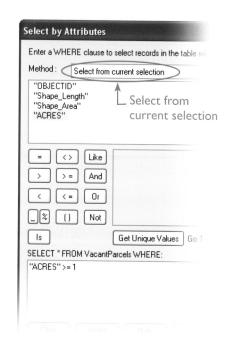

Figure 6-24

ⓙ Open the *VacantParcels* attribute table.

You should have just 164 of 15,802 parcels selected.

2) Subselect parcels one acre or larger. Of the 164 selected parcels, we want to keep only those that are at least one acre. We'll do this with an attribute query on the ACRES field.

ⓐ At the top of the table, click the Table Options button 🔡 and choose Select By Attributes.

ⓑ In the Select By Attributes dialog box, click the Method drop-down arrow and choose "Select from current selection."

This restricts the scope of the query to records that are already selected.

ⓒ Build the following expression: "ACRES" >= 1.

ⓓ Check your dialog box against Figure 6-24. Click Apply, then close the Select By Attributes dialog box.

We're down to seven selected records: just seven parcels meet all our conditions.

ⓔ In the Table of Contents, right-click *VacantParcels* and choose Selection → Zoom To Selected Features.

ⓕ In the attribute table, show the selected records ▤ (Figure 6-25).

ⓖ Close the attribute table.

At this scale, it's hard to see the selected parcels on the map. We'll take a closer look in the next step.

Figure 6-25

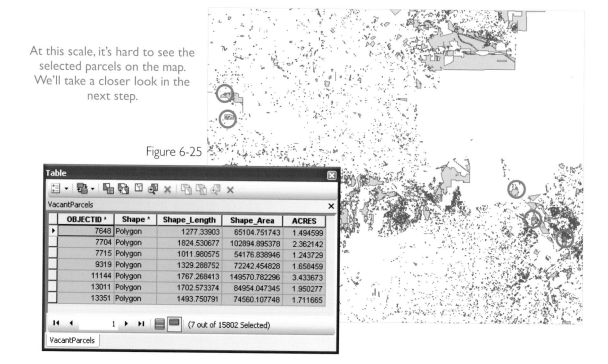

3) Export selected features. These seven parcels are the critical ones we want, they represent the culmination of the process, and we want to save them as a feature class. Again, we'll use the convenient Export Data command.

Ⓐ In the Table of Contents, right-click *VacantParcels* and choose Data ➔ Export Data.

Ⓑ Highlight and delete the default output feature class name (Export_Output). Replace it with **SevenSites**.

Ⓒ Check your settings against Figure 6-26, then click OK.

Ⓓ Click Yes to add the exported data as a layer.

4) Inspect the candidates. Now we'll zoom in and take a close look at the seven sites against basemap imagery.

Ⓐ Add the *Imagery* basemap layer.

▷ If you get a geographic coordinate systems warning, close it.

Ⓑ In the Table of Contents, remove *VacantParcels*. Turn off all layers except *SevenSites* and *Basemap*.

Ⓒ In the Table of Contents, open the Symbol Selector for the *SevenSites* layer.

Ⓓ Set the fill color to No Color and the outline color to Medium Apple (Figure 6-27). Make the outline width 2. Click OK.

Ⓔ Open the attribute table for the *SevenSites* layer and move it away from the map.

Ⓕ In the table, right-click the gray box next to the first record and choose Zoom To (Figure 6-28).

The map zooms to the first candidate parcel (Figure 6-29).

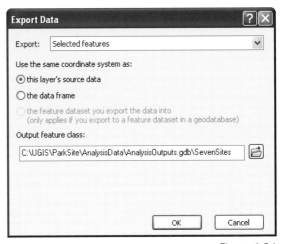

Figure 6-26

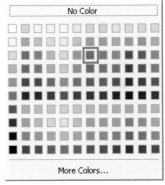

Figure 6-27

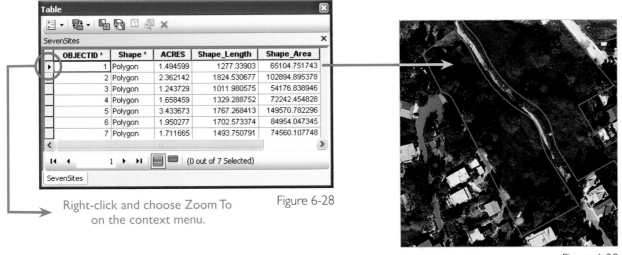

Right-click and choose Zoom To on the context menu.

Figure 6-28

Figure 6-29

G On the Standard toolbar, click the Scale Box drop-down arrow and choose 1:10,000.

H If necessary, pan to the southwest so you can see the river.

Notice that three of our seven sites are right next to each other. (We'll come back to this point in the next step.) Notice also that these sites aren't especially close to the river.

I In the Table of Contents, turn on the *ProximityZone* layer and the *Parks* layer.

J Optionally, symbolize the *Parks* layer in a shade of green (Figure 6-30).

Figure 6-30

The view confirms that the sites are near the edge of the proximity zone. Even more interesting is that another park is closer to the river on a direct line. That's why human interpretation of analysis results is so important. We considered proximity to existing parks, but it didn't occur to us that a park might be more than a quarter mile from a candidate site and yet closer to the river! It doesn't invalidate our candidates, but it's something to think about.

K Turn off the *ProximityZone* and *Parks* layers.

L In the table, right-click the gray box next to the fourth record and choose Zoom To.

M Zoom to a map scale of **1:2,500** (Figure 6-31).

Figure 6-31

This site is right next to the river, which is ideal. Notice a couple of other things. First, the river bottom is natural here, which could make this site more scenic than others. Our analysis doesn't take aesthetic considerations into account, but the views, sounds, and smells of a park affect the people who use it.

On the other hand, the site is also adjacent to a freeway. This has an aesthetic impact and probably a health impact, as well: should kids be exposed to all that car and truck exhaust? The freeway is also a barrier to accessibility. We haven't yet done the part of the analysis that considers the total population served, but our formula uses a simple distance buffer that doesn't consider freeways or other barriers (see Appendix A).

Figure 6-32

Ⓝ Zoom to the remaining sites (Figure 6-32) and investigate them:

• Zoom and pan to see the site in relation to the river.
• Turn other map layers on and off.
• Make observations that add context to the analysis.

Ⓞ When you're finished, turn off all layers except *SevenSites* and *Basemap*.

▷ Leave the attribute table open.

This site is next to railroad tracks.

5) Aggregate sites. Our first three sites are separated by narrow gaps. These are probably unpaved easements, or rights-of-way, intended to become streets in the event of development. Because the sites are so close together, and so much alike, it seems odd to present them as distinct choices. We can't dissolve them—they don't share boundaries—but we can achieve a similar effect with a tool called Aggregate Polygons. This tool merges nearby polygon features on the basis of a specified distance value. We'll make measurements to get a sense of what this value should be, then aggregate the sites to give the city council a "large park" option.

This site has an irregular shape.

Ⓐ In the attribute table, zoom to the first record, then close the table.

Ⓑ Zoom out a little and pan until the three nearby parcels more or less fill your screen.

Ⓒ On the Tools toolbar, click the Measure tool ⟐.

Ⓓ On the Measure dialog box, make sure the Measure Line tool is selected (Figure 6-33).

Ⓔ Click the Choose Units drop-down arrow and choose Distance → Feet.

This site is partially outside the proximity zone.

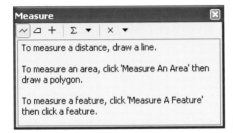

Figure 6-33

F Draw lines to measure the distances at a few different places along the gaps between the parcels (Figure 6-34).

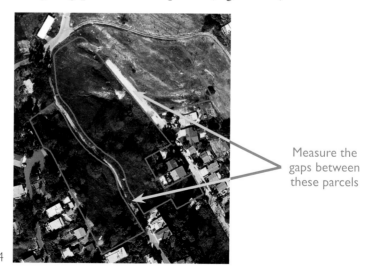

Measure the gaps between these parcels

Figure 6-34

You should get measurements of about 30 feet.

G Close the Measure dialog box.

Before we go ahead with the aggregation, let's acknowledge the assumptions we've made about land use: that any vacant parcel is available regardless of ownership, that it's okay to combine adjacent parcels, and now that it's okay to combine nearly adjacent parcels. Any of these assumptions may turn out to be problematic in reality, but it seems reasonable to suppose that the acquisition of land for a public park might require buying multiple parcels.

Unlike the other tools we're using in this lesson, Aggregate Polygons is found under Cartography Tools in the Generalization toolset. The main purpose of generalization tools is to simplify or refine the appearance of features for small-scale map display—our use of the tool in this context is a little bit unorthodox.

H Open the Search window. If necessary, click Tools near the top. Delete the existing search term, if any, then start to type **aggregate**.

I On the drop-down list of tools, choose "aggregate polygons (cartography)" to locate it.

J In the list of results, click Aggregate Polygons (Cartography) to open the tool.

▷ Be careful not to choose Aggregate Points.

K In the Aggregate Polygons dialog box, set the Input Features to *SevenSites*.

L Name the Output Feature Class **FiveSites**.

Ⓜ Enter an Aggregation Distance of **50** and make sure the units are Feet.

We're allowing a little margin of error.

Ⓝ Check your settings against Figure 6-35 and click OK.

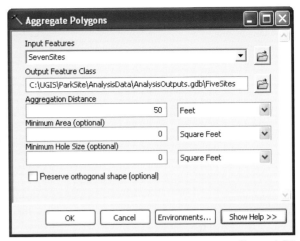

Figure 6-35

When the tool finishes, a layer called *FiveSites* and a table called *FiveSites_Tbl* are added to the Table of Contents. On the map, you can see the new, aggregated polygon on top of the three original sites (Figure 6-36).

The Table of Contents has switched to the List By Source view to show the table.

Ⓞ In the Table of Contents, open *FiveSites_Tbl*.

This backlink table shows which of the features in *SevenSites* were aggregated. We might need it under other circumstances, but not here.

Ⓟ Close *FiveSites_Tbl* and remove it from the Table of Contents.

Ⓠ At the top of the Table of Contents, click the List by Drawing Order button 🔁.

Ⓡ Open the *FiveSites* attribute table.

We're now down to five candidates. (Only the three sites in the view were aggregated because none of the others were near each other.) Note that we've lost the ACRES attribute from the *SevenSites* layer. A consequence of aggregating features, as with dissolving them, is the loss of attributes.

6) **Calculate acreage.** To restore acreage values, we'll add a field and calculate its geometry.

Figure 6-36

Ⓐ In the *FiveSites* attribute table, click the Table Options button 🗒 and choose Add Field.

Ⓑ Name the field **ACRES** and set the Type to Double (Figure 6-37), then click OK.

Ⓒ In the attribute table, right-click the ACRES field and choose Calculate Geometry.

▷ You won't get a warning if you turned warnings off in Lesson 4 (page 153). If you do, click Yes to continue.

Ⓓ In the Calculate Geometry dialog box, set the units to Acres US [ac] and accept the other defaults. Click OK.

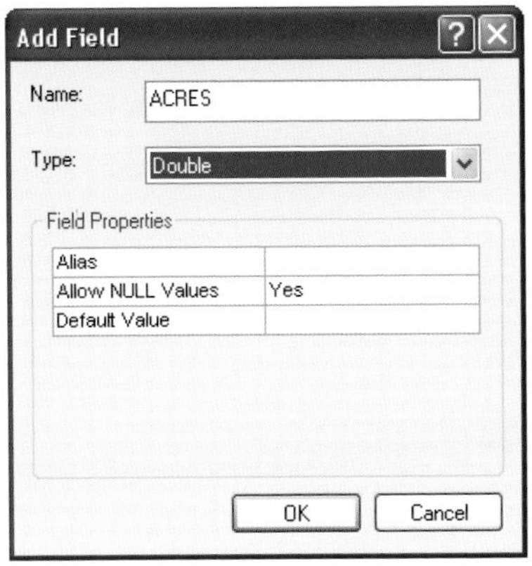

Figure 6-37

The field is populated with acreage values (Figure 6-38).

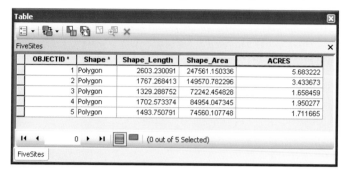

OBJECTID *	Shape *	Shape_Length	Shape_Area	ACRES
1	Polygon	2603.230091	247561.150336	5.683222
2	Polygon	1767.268413	149570.782296	3.433673
3	Polygon	1329.288752	72242.454828	1.658459
4	Polygon	1702.573374	84954.047345	1.950277
5	Polygon	1493.750791	74560.107748	1.711665

Figure 6-38

E Close the Attribute table.

At this point, there are two items remaining from our list of analysis criteria: we want the new park to serve the maximum number of people and we want it to be as close to the river as possible.

7) **Define park access zones.** In Lesson 2, we defined park accessibility in terms of a quarter-mile distance. This is an oversimplification of accessibility, but it has the virtue of being easy to analyze: we just have to buffer the candidate sites and then count the population inside them.

A In the Table of Contents, turn off the *SevenSites* layer.

B Right-click the *FiveSites* layer and choose Zoom to Layer.

You probably can't see the sites anymore, but that's okay.

C From the main menu, choose Geoprocessing→Buffer.

D In the Buffer dialog box, set the Input Features to *FiveSites*.

E Name the Output Feature Class **ParkAccessZones**.

F In the Linear Unit box, type **0.25** and set the units to Miles.

G Double-check your settings against Figure 6-39, then click OK.

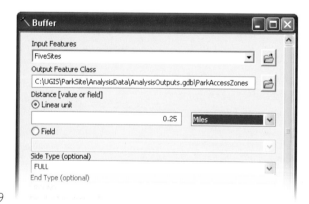

Figure 6-39

When the tool finishes, the *ParkAccessZones* layer is added to the map. You can see the five roughly circular features representing the quarter-mile access zones around the candidates (Figure 6-40).

8) Add block centroids to the map. We need to find the population within each of these access zones. The basic approach of counting the block centroids in each zone and summing their populations was introduced in Lesson 2 (page 80) and sketched in the Analysis Plan at the beginning of this lesson. For another way, see *Apportioning attribute values* on page 223.

Figure 6-40

How do we do it? With a spatial query, we could select block centroids within park access zones. Then, in the *BlockCentroids* table, we could right-click the POP2000 field, choose Statistics, and see the sum of the selected records. It would mean doing five operations (each access zone has to be queried separately), but that's not a big deal. A bigger drawback is that this method returns the answer in a message box— we want to write it to a table.

We can do that using the Spatial Join tool. You might think of a spatial join as a spatial query plus a table join: attributes from one table are joined to another on the basis of a spatial relationship between layers, rather than a common attribute.

Let's get a visual sense of what we're going to accomplish.

Figure 6-41

Ⓐ Turn off the *Basemap* layer.

Ⓑ Add *BlockCentroids* to the map from ReadyData.gdb.

Ⓒ Drag *ParkAccessZones* to the top of the Table of Contents.

Ⓓ Symbolize *ParkAccessZones* with a fill color of No Color. Make the outline width 2 and choose any outline color you want.

Ⓔ Zoom in on one of the access zones (Figure 6-41).

Ⓕ Open the *BlockCentroids* attribute table (Figure 6-42).

The POP2000 field values will be summed for block centroids within each access zone and written to a new feature class attribute table.

Ⓖ Close the *BlockCentroids* table.

Figure 6-42

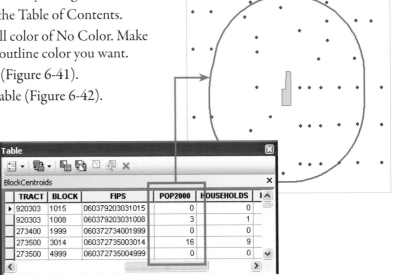

TRACT	BLOCK	FIPS	POP2000	HOUSEHOLDS	
920303	1015	060379203031015	0	0	
920303	1008	060379203031008	3	1	
273400	1999	060372734001999	0	0	
273500	3014	060372735003014	16	9	
273500	4999	060372735004999	0	0	

Figure 6-43

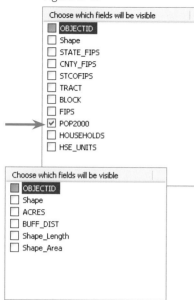

Figure 6-44

Figure 6-45

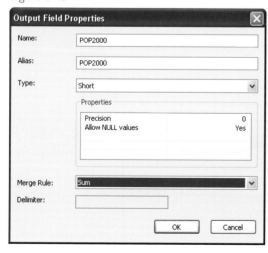

The only *BlockCentroids* attribute we need is POP2000. We'll turn the other fields off so they're not carried forward.

Ⓗ Open the layer properties for *BlockCentroids* and click the Fields tab.

Ⓘ Click the Turn all fields off button ⊡.

Ⓙ In the list of fields, check the POP2000 box to turn it back on (Figure 6-43), then click OK.

We don't need any of the attributes from the *ParkAccessZones* layer.

Ⓚ Open the layer properties for *ParkAccessZones*. On the Fields tab, click the Turn all fields off button (Figure 6-44). Click OK.

9) Spatially join block centroids to park access zones. We're ready to do the spatial join. Unlike a table join, a spatial join isn't virtual—the output is a new feature class.

Ⓐ Open the Search window. Highlight and delete the existing search term, then start to type **spatial join**.

Ⓑ On the drop-down list of tools, choose "spatial join (analysis)" to locate it.

Ⓒ In the list of results, click Spatial Join (Analysis) to open the tool.

Ⓓ In the Spatial Join dialog box, set the Target Features to *ParkAccessZones*.

Ⓔ Set the Join Features to *BlockCentroids*.

Ⓕ Accept the default Output Feature Class name of *ParkAccessZones_SpatialJoin*.

In the Field Map area, the only fields are POP2000 and the required length and area fields. Note that POP2000 has a Double field type, which isn't really necessary for countable data.

Ⓖ In the Field Map area, right-click POP2000 and choose Properties.

Ⓗ In the Output Field Properties dialog box, click the Type drop-down arrow and choose Short.

Ⓘ Click the Merge Rule drop-down arrow and choose Sum (Figure 6-45), then click OK.

The merge rule setting is crucial. It tells ArcMap that for each target feature (access zone) we want to sum the population values of the join features (block centroids).

Ⓙ Below the Field Map area, click the Match Option drop-down arrow and choose CONTAINS.

▷ You may need to lengthen the dialog box to see all the parameters.

Ⓚ Compare your dialog box to Figure 6-46, then click OK.

When the tool finishes, a *ParkAccessZones_SpatialJoin* layer is added to the map.

Ⓛ Open the attribute table for the *ParkAccessZones_SpatialJoin* layer (Figure 6-47).

The POP2000 field has the total population within a quarter mile of each site. We've written these values to a table, which is what we wanted, but we haven't written them to the *FiveSites* table, which is really *where* we want them. We'll take care of this in just a little bit.

Ⓜ Close the table.

10) Calculate distance to the river. Our last requirement is to figure out the distance from each site to the LA River. We could use the Measure tool to do that, of course, but we'd face the same situation of writing the results to a table. Also, we might not be sure of making the shortest possible measurement. The Near tool will make these calculations for us automatically.

Ⓐ Open the Search window. Highlight and delete the existing search term, then type **near**.

Ⓑ On the drop-down list of tools, choose "near (analysis)" to locate it.

Ⓒ In the list of results, click Near (Analysis) to open the tool.

Figure 6-46

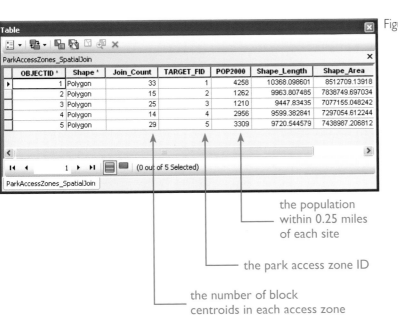

Figure 6-47

the population within 0.25 miles of each site

the park access zone ID

the number of block centroids in each access zone

D In the Near dialog box, set the Input Features to *FiveSites*.

E Set the Near Features to *LARiver*.

Note that you can add multiple layers to the list of Near features. Typically, the tool is used to discover which feature, in one or more layers, is closest to a feature of interest. In our case, we just want the distance measurement.

The Near tool doesn't create a new feature class. Instead, it will add a couple of fields to the *FiveSites* table.

F Check your settings against Figure 6-48, then click OK.

G When the tool finishes, open the *FiveSites* layer attribute table and scroll to the right (Figure 6-49).

Figure 6-48

Figure 6-49

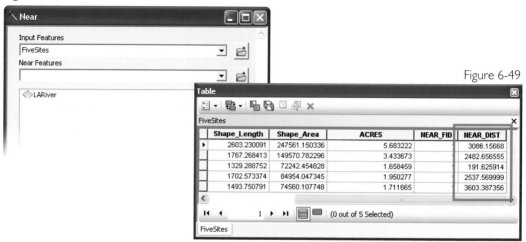

The NEAR_DIST field stores the distance, in feet, of each site to the river.

H Close the table.

11) Add the park-accessible population to *FiveSites*. It would be nice to store all the relevant site-selection attributes in a single table with the candidate sites:

- Acreage
- Distance to the river
- Total population within a quarter-mile
- Demographic variables

That will make it easy to evaluate and compare the sites at a glance—for us and for anyone we share the data with. We already have the acreage and distance attributes in the *FiveSites* table. In this step, we'll add the population totals that we got from the spatial join in Step 9.

We could do this with another spatial join, but we'll go back to the Identity tool as a matter of preference. As we've done before, we'll filter out the attributes we don't need.

Ⓐ Open the layer properties for *ParkAccessZones_SpatialJoin*.

Ⓑ On the Fields tab, click the Turn all fields off button ⊟ .

Ⓒ Check the POP2000 box to turn it back on (Figure 6-50), then click OK.

Ⓓ Open the Identity (analysis) tool.

Ⓔ For the Input Features, choose *FiveSites*.

Ⓕ For the Identity Features, choose *ParkAccessZones_SpatialJoin*.

Ⓖ Accept the default Output Feature Class name of *FiveSites_Identity*.

Ⓗ Check your settings against Figure 6-51, then click OK.

When the tool finishes, the *FiveSites_Identity* layer is added to the map.

Ⓘ Open the *FiveSites_Identity* layer attribute table and scroll to the right (Figure 6-52).

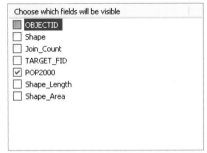

Figure 6-50

Figure 6-51

Figure 6-52

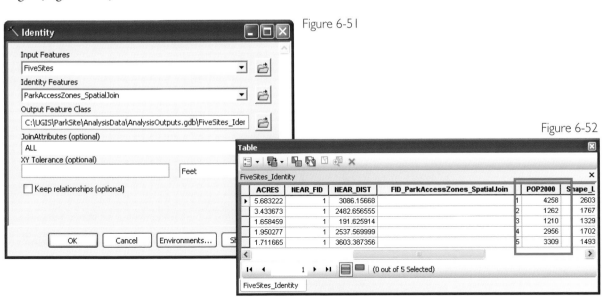

The POP2000 values (representing population within a quarter mile) have been brought in.

Ⓙ Close the table.

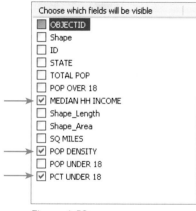

Figure 6-53

12) Add the demographic attributes to *FiveSites_Identity*. We'll use Identity once more to get the demographic attributes into the table. These attributes are found in the *BlockGroups* layer.

- **Ⓐ** Open the layer properties for *BlockGroups*.
- **Ⓑ** On the Fields tab, click the Turn all fields off button.
- **Ⓒ** Check the following fields to turn them back on:
  - MEDIAN HH INCOME
  - POP DENSITY
  - PCT UNDER 18

Check your settings against Figure 6-53, then click OK.

- **Ⓓ** Open the Identity (analysis) tool.
- **Ⓔ** For the Input Features, choose *FiveSites_Identity*.
- **Ⓕ** For the Identity Features, choose *BlockGroups*.
- **Ⓖ** Rename the Output Feature Class **RecommendedSites**.
- **Ⓗ** Check your settings against Figure 6-54, then click OK.

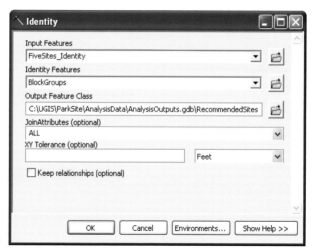

Figure 6-54

When the tool finishes, the *RecommendedSites* layer is added to the map.

- **Ⓘ** Open the *RecommendedSites* layer attribute table and scroll across it.

It has all the attributes of interest, plus several ID attributes (backlinks to features in other tables) that we don't need. In the next exercise, we'll format this table to make it presentable.

- **Ⓙ** Close the table.
- **Ⓚ** Save the map.
- **Ⓛ** If you are continuing to the next exercise now, leave ArcMap open. Otherwise, exit ArcMap.

Apportioning attribute values

Splitting feature geometry has implications for attribute values. In Example 1, a census block group is split by a buffer in an overlay operation, resulting in two output features. By default, ArcMap copies the attribute values of the input feature to both output features. That's okay for POPDENSITY, which is already a ratio: the value applies to the parts as well as to the whole. It's not okay for TOTALPOP, a count value, because it doubles the population, as shown.

In Example 2, we want to count the population inside the buffer, which covers parts of three block groups. A reasonable approach is to analyze the percentage of each block group's area that falls inside the buffer and then assign the same percentage of its population value. In this case, 1% of block group 1's area is inside the buffer, so its contribution to the buffer population should be 1% of 1,755. Likewise for block group 2 on the far right (21% of 4,445) and block group 3 in the middle (38% of 6,722).

The examples are variations of the same problem, which can be solved with a tool called Make Feature Layer. This tool makes a layer from a feature class (essentially the same thing that happens automatically when you add data to ArcMap). It has a very useful parameter, however: a "ratio policy" check box for each attribute in the input feature class. When you geoprocess a layer made with this tool, the ratio policy is applied to any split features, dividing their attribute values according to area.

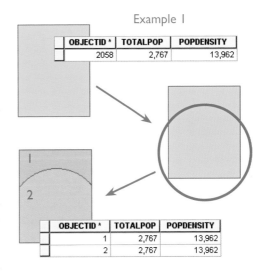

Example 1

OBJECTID *	TOTALPOP	POPDENSITY
2058	2,767	13,962

OBJECTID *	TOTALPOP	POPDENSITY
1	2,767	13,962
2	2,767	13,962

Example 2

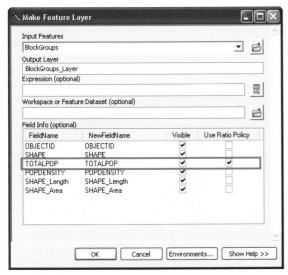

OBJECTID	TOTALPOP
1	1,755
2	4,445
3	6,722

The "Use Ratio Policy" box is checked for the TOTALPOP field.

Make Feature Layer is run on the *BlockGroups* feature class. Subsequent overlay operations give desired results.

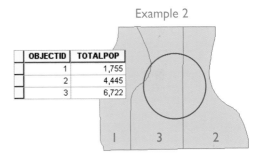

Example 1 solution

OBJECTID *	TOTALPOP	POPDENSITY
1	762	13,962
2	2,005	13,962

Example 2 solution

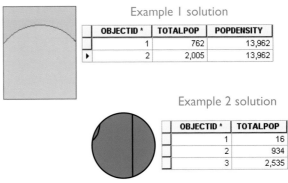

OBJECTID *	TOTALPOP
1	16
2	934
3	2,535

The final step to get the total population of the buffer area would be to use the Dissolve tool with TOTALPOP as a statistics field.

Exercise 6d: Clean up the map and geodatabase

Analysis projects tend to leave clutter behind. It's normal to find your working map and geodatabase filled with a mixture of data you want to preserve and data you can now discard. It's tempting to leave the mess behind, but sooner or later someone will return to the scene and wish that you'd taken the time to put things in order. In this exercise, we'll simplify our map document, which contains several unnecessary layers. We'll also format the *FiveSites* layer attribute table and save it as a layer file. Finally, we'll remove intermediate data from the geodatabase and add some metadata to the datasets we want to keep.

1)   Clean up the map document. This isn't the map document we'll use for our final layout, but we still want it to be intelligible. With that in mind, we'll keep just a few important layers in it.

ⓐ If necessary, start ArcMap and open Lesson6.

ⓑ In the Table of Contents, remove all layers except the four listed below. Put the layers in the top-to-bottom order shown:

- *RecommendedSites*
- *LARiver*
- *LARiverBuffer*
- *Basemap*

ⓒ Turn all four layers on and zoom to the *LARiverBuffer* layer.

ⓓ Symbolize the *LARiver* layer with the color Big Sky Blue (Figure 6-55) and a width of 3.

ⓔ Symbolize *RecommendedSites* with a fill color of No Color, an outline width of 2, and an outline color of Medium Apple.

ⓕ Symbolize *LARiverBuffer* with a fill color of No Color, an outline width of **1.5**, and an outline color of Medium Lilac.

Your map should look like Figure 6-56.

Figure 6-55

Big Sky Blue

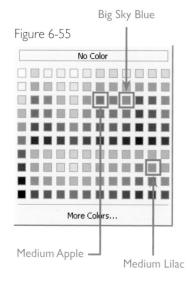

Medium Apple

Medium Lilac

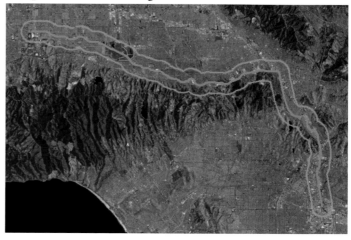

Figure 6-56

2) Add layer descriptions. When you add a layer to a map, a description—taken from the item description of the source data—is added to the layer properties. This description explains what the layer represents. Layer descriptions are helpful to anyone using the map document and are required if you share layers as packages. The *LARiver* and *Basemap* layers already have descriptions; the other two, created during geoprocessing, do not.

Ⓐ Open the layer properties for *RecommendedSites* and click the General tab.

Ⓑ In the Description box, type: **These are the five sites recommended for a new park near the Los Angeles River.** (Figure 6-57) Click OK.

Ⓒ Open the layer properties for *LARiverBuffer*. In the Description box, type: **This is a 3/4 mile buffer around the Los Angeles River.** (Figure 6-58) Click OK.

Figure 6-57

3) Delete unnecessary fields. In the last exercise, we got all the important attributes into the *RecommendedSites* table. We also have several FID attributes that we don't need to maintain.

Ⓐ Open the *RecommendedSites* attribute table.

Ⓑ In the table, right-click the FID_FIVESITES_IDENTITY field heading (Figure 6-59) and choose Delete Field.

Ⓒ Click Yes on the prompt to confirm the field deletion.

The field is deleted from the table.

Figure 6-58

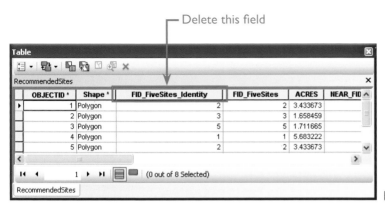

Delete this field

Be careful when you delete a field because you can't undo it.

Figure 6-59

D In the same way, delete all the remaining FID fields from the table. They are:

- FID_FIVESITES
- NEAR_FID
- FID_PARKACCESSZONES_SPATIALJOIN
- FID_BLOCKGROUPS

The table should look like Figure 6-60.

Figure 6-60

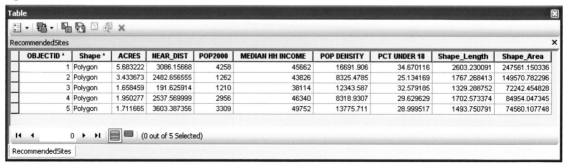

4) **Add a field.** We're going to add an identifier field that we can manage. We'll have a reason to use this field in Lesson 8.

A Click the Table Options button 🗏 and choose Add Field.

B Name the field **SiteID** and accept the default Short Integer field data type.

C Check your settings against Figure 6-61, then click OK.

D Scroll to the end of the table. Right-click the SITEID field heading and choose Field Calculator.

E In the Field Calculator dialog box, double-click OBJECTID in the list of fields to add it to the expression box.

This is a simple expression that will assign the same values of 1 to 5 to the new field. The difference is that SITEID is a field we can manage, while OBJECTID is not.

F Check your settings against Figure 6-62, then click OK.

Figure 6-61

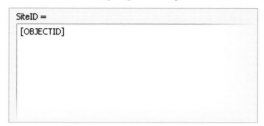

Figure 6-62

The field is populated with the values.

G Leave the table window open, but move it out of the way.

5) Format fields. Most of the fields in the table should be aliased or formatted in other ways to make them more readable. We'll go through the steps in detail for one field and let you do the rest with general instructions.

Ⓐ Open the layer properties for *RecommendedSites* and click the Fields tab.

Ⓑ In the list of fields, uncheck SHAPE_LENGTH and SHAPE_AREA to hide them.

Ⓒ At the bottom of the list, click SITEID to highlight it.

Ⓓ In the row of tools at the top, click the little drop-down arrow next to the Move Up button ↑▾ and choose Move To Top.

Ⓔ Click the Move Down button ↓▾ twice to move the field directly underneath SHAPE (Figure 6-63).

Moving this field affects the layer only. In the feature class, fields are stored in the order in which they're added, and this order can't be changed.

Ⓕ Click the PCT UNDER 18 field near the bottom of the list to highlight it.

Ⓖ In the box on the right, under Appearance, click the word "Alias" to highlight it. Replace the current alias with **Percent Under 18 Years**.

Ⓗ Two lines down, click Number Format to highlight it.

The next three steps are shown as a sequence in Figure 6-64.

Arrange the display order of fields with these buttons

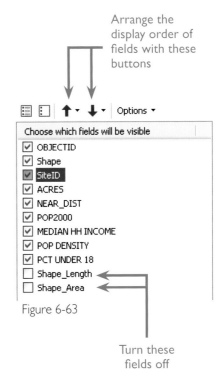

Figure 6-63

Turn these fields off

Click to format as %

Figure 6-64

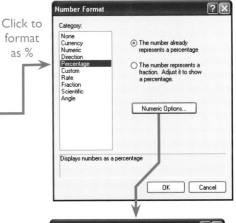

Change to zero decimal places

Ⓘ Click the button that appears on the right to open the Number Format dialog box.

Ⓙ In the Number Format dialog box, click the Percentage category, then click Numeric Options.

Ⓚ Change the number of decimal places to 0.

Ⓛ Click OK on the Numeric Options and Number Format dialog boxes.

Ⓜ Click Apply to apply your changes so far.

N Format the remaining fields according to the following table. Depending on the field, you may have to

- Change or add an alias.
- Change the number format.
- Change the number of decimal places.
- Check the box to show thousands separators.

Field	Alias	Number Format
ACRES	Acres	Numeric (one decimal place)
NEAR_DIST	Feet to LA River	Numeric (zero decimal places; show thousands separators)
POP2000	1/4 Mile Population	Numeric (show thousands separators)
MEDIAN HH INCOME	Median Household Income	Currency
POP DENSITY	People per Square Mile	Numeric (zero decimal places; show thousands separators)

O Click OK to close the Layer Properties.

P Widen fields as needed by placing the mouse pointer between field headings and dragging in either direction.

Your table should look like Figure 6-65.

Figure 6-65

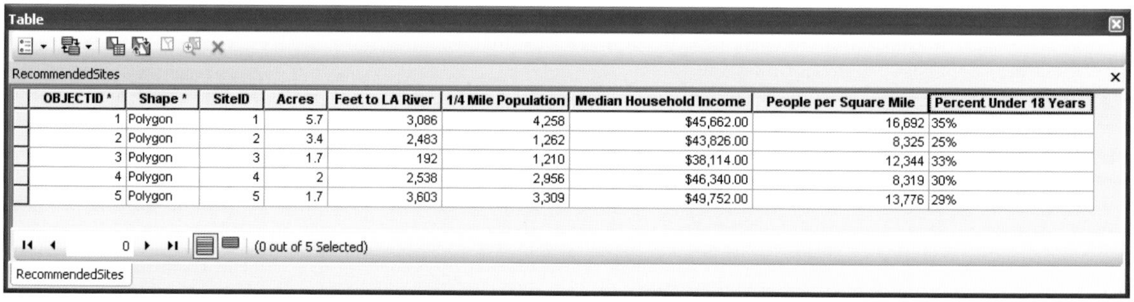

OBJECTID *	Shape *	SiteID	Acres	Feet to LA River	1/4 Mile Population	Median Household Income	People per Square Mile	Percent Under 18 Years
1	Polygon	1	5.7	3,086	4,258	$45,662.00	16,692	35%
2	Polygon	2	3.4	2,483	1,262	$43,826.00	8,325	25%
3	Polygon	3	1.7	192	1,210	$38,114.00	12,344	33%
4	Polygon	4	2	2,538	2,956	$46,340.00	8,319	30%
5	Polygon	5	1.7	3,603	3,309	$49,752.00	13,776	29%

Q If necessary, open the layer properties and make any needed changes.

R Close the attribute table.

6) Save the layer as a layer file. All these format settings are layer properties. If we want them to be available in other map documents, we need to save *RecommendedSites* as a layer file, as we did with the city boundary of Los Angeles in Lesson 1.

A In the Table of Contents, right-click *RecommendedSites* and choose Save As Layer File.

B In the Save Layer dialog box, navigate if necessary to Home - ParkSite\MapsAndMore.

C Accept the default name *RecommendedSites.lyr* (Figure 6-66) and click Save.

7) Set map document properties. Maps, like datasets, can have metadata. By filling out a few fields, you can supply some background information about your map and make it searchable.

A From the main menu, choose File→Map Document Properties.

B In the Map Document Properties dialog box, for the Title, type: `Los Angeles River Park Site Analysis`.

C For the Summary, type: `Map of recommended sites for park development near the Los Angeles River.`

D For the Description, type: `These are the five sites recommended for a new park near the Los Angeles River.`

E For the Author, type your name.

F For the Credits, type: `City of Los Angeles, ESRI.`

G For the Tags, type: `parks, Los Angeles River, Los Angeles.`

▷ Include the commas to separate the tags.

H Near the bottom of the dialog box, check the box to store relative pathnames to data sources.

This setting allows you to move the map document and its data sources to different locations without disruption, as long as the layer paths to source data maintain the same directory structure relationship.

I Click Make Thumbnail.

J Compare your settings to Figure 6-67, then click OK.

K Save the map.

L In the Catalog window, in the Home folder, right-click *Lesson6.mxd* and choose Item Description.

The properties you set appear in the Item Description for the map document. (If you don't see the thumbnail graphic, right-click the Home folder in the Catalog window and choose Refresh.) Note that the Preview tab allows you to navigate the map.

M Close the Item Description window.

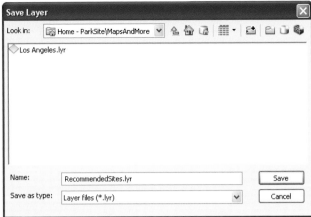

Figure 6-66

Figure 6-67

8) Clean up the geodatabase. In this step, we'll delete intermediate data from the geodatabase and update the item descriptions for the two feature classes that we're keeping.

Ⓐ In the Catalog window, expand AnalysisOutputs.gdb and click on it.

Ⓑ If necessary, click the Show Next View button 🔲 to split the Catalog window into a top and bottom pane.

The geodatabase should contain eleven feature classes and one table, as shown in Figure 6-68.

Ⓒ In the bottom pane of the Catalog window, hold the Control key and click each dataset in AnalysisOutputs to select it, except these two:

- *LARiverBuffer*
- *RecommendedSites*

Ⓓ Compare your Catalog window to Figure 6-69, then right-click any of the selected datasets and choose Delete.

Figure 6-68

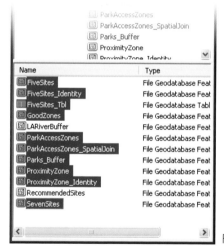

Figure 6-69

Ⓔ Click Yes to delete the datasets.

Now we'll update the Item Descriptions for the two remaining feature classes. The ones they have now have been copied from their geoprocessing inputs.

Ⓕ In the Catalog window, right-click the *LARiverBuffer* feature class and choose Item Description.

Ⓖ In the Item Description window, click the Preview tab and click the Create Thumbnail button 🔲 .

Ⓗ Click the Description tab. In the row of buttons at the top, click Edit.

ⓘ In the Summary box, highlight and delete the existing text. Replace it with this: `This dataset represents the area under consideration for a new park near the Los Angeles River.`

ⓙ In the Description box, highlight and delete the existing text. Replace it with this: `This dataset is a 3/4 mile buffer around the Los Angeles River.`

ⓚ In the Credits box, highlight and delete the existing text. Replace it with this: `City of Los Angeles, ESRI.`

ⓛ At the top of the window, click Save (Figure 6-70).

ⓜ Leave the Item Description window open. In the Catalog window, click *RecommendedSites.*

The Item Description window updates.

ⓝ Click the Preview tab and click the Create Thumbnail button ▦.

ⓞ Click the Description tab and click Edit.

ⓟ In the Summary box, highlight and delete the existing text. Replace it with this: `These are the five sites recommended for a new park near the Los Angeles River.`

ⓠ In the Description box, highlight and delete the existing text. Replace it with this: `This dataset is a subset of vacant land parcels one acre or larger near the Los Angeles River.`

ⓡ In the Credits box, replace the existing text with this: `City of Los Angeles, ESRI.`

ⓢ Click Save and close the Item Description window.

ⓣ Save the map.

Figure 6-70

Exercise 6e: Evaluate results

The recommended sites meet our criteria and look good on basemap imagery, but ultimately, there's no substitute for on-site inspection. How do these candidate locations really look in the context of their surroundings? If you had the opportunity, it would be worthwhile to visit the sites and record your impressions. In this exercise, we've included some photographs of the sites to help give a sense of that experience.

1) **Compare your guesses from Lesson 1.** In Lesson 1 you made some guesses about likely areas for parks based on a partial look at the data. Let's see how those turned out.

Ⓐ In the Catalog window, under the Home folder, drag and drop *Lesson1Predictions.tif* onto the map.

▷ Close the Geographic Coordinate Systems Warning if you get it.

▷ If you don't have this file, you can download it from the Lesson 1 results at the Understanding GIS Resource Center.

Ⓑ Zoom to the *Lesson1Predictions.tif* layer (Figure 6-71).

The layer is partially transparent because you saved it this way in Lesson 1.

Figure 6-71

The exported map is a raster dataset. Its GeoTIFF tags contain the coordinate system information ArcMap needs to align it with the other layers in the map.

How did your guesses turn out? You can zoom and pan as needed to get a better look.

Ⓒ When you're finished, turn off the *Lesson1Predictions.tif* layer.

Converting graphics to features

Another way to make the comparison is to convert the black ellipses from graphic elements to features. Here are the steps:

- Open Lesson1.mxd.
- On the main menu, click Edit and choose Select All Elements.
- Add the Draw toolbar.
- On the Draw toolbar, click the Drawing drop-down menu and choose Convert Graphics to Features.
- Save the feature class in AnalysisOutputs.gdb.
- Add the new feature class to the Lesson6 map.

2) Examine candidates. We'll take a short virtual tour of the five candidate sites.

ⓐ Open the *RecommendedSites* attribute table and zoom to Site 1 (Figure 6-72), then zoom to Site 2 (Figure 6-73).

Figure 6-72

Site 1 has good demographics but is far from the river.

The site is hilly with limited road access.

Figure 6-73

Site 2 has comparatively weak demographics. It's about a half mile from the river, placing it in the middle of the candidates.

The site is now being used informally as a soccer field.

Figure 6-74

Site 3 is adjacent to the river. It has the lowest median household income and the second highest percentage of children.

The site is also right next to a freeway.

Zoom to Site 3 (Figure 6-74), then zoom to Site 4 (Figure 6-75).

Figure 6-75

Site 4 has relatively poor demographics. Like Site 2, it's about half a mile from the river. It has a developed residential parcel jutting into it.

The site is apparently used for storage (above), and is next to high-voltage power lines (below).

c Zoom to Site 5 (Figure 6-76).

Figure 6-76

Site 5 has mixed demographics. It is the farthest site from the river, actually crossing the 3/4-mile buffer (purple line).

The site is easily accessed from streets.

d Close the attribute table and zoom to the *LARiverBuffer* layer.

e Save the map and exit ArcMap.

This wraps up our analytical work. We started with a set of (fictional) guidelines from the city council. Those guidelines were vague in many cases, so we ended up supplying our own working definitions. Obviously, these could be questioned. The guidelines also excluded many factors that might be important: whether the land is actually available, its environmental condition, barriers to access such as freeways, and several more. A number of caveats to the analysis are discussed in Appendix A. Finally, what the council did not give us—and we have no way to evaluate on our own—is the relative importance of the criteria. We can't rank our candidates because we don't know, for example, if low household income outweighs a high percentage of children in making a site desirable, and if so, by how much. We have to present the options neutrally because we don't know what the council's priorities are. (They may not know them either until they begin to study the results.)

Suppose somebody asks you how the results would change if you set the income requirement lower or the percentage of children requirement higher? How would you answer simple questions like that without rerunning the entire analysis from scratch? Read on to Lesson 7 for the solution.

Lesson

7 Automate the analysis

SUPPOSE UNEXPECTED CHANGES occur after you've finished the analysis. Maybe an updated table of vacant parcels becomes available. Maybe after seeing the results, the City Council wants to add or remove criteria, or see how a change in threshold values affects the outcome. Of course, you could redo the analysis, but even a very small change—especially if it comes at the beginning of the process—could make for a lot of tedious work. This is the moment to introduce ModelBuilder™. ModelBuilder can encapsulate your entire workflow in a single geoprocessing tool that runs with the push of a button.

A model looks like a flowchart or schematic diagram that shows the steps for manufacturing a product. Inputs, outputs, and processes are represented by geometrical shapes. Arrows connect the shapes to show the sequence of operations. For an example, see the topic *Understanding a model* on page 240.

A model is more than just a visual aid; it's also an engine. The schematic elements representing datasets and tools are connected, so that running a model sets in motion a chain reaction of geoprocessing operations. In our case, the model will basically repeat the analysis tasks we did in Lesson 6.

There are lots of good reasons to build a model from our analysis. First, being a diagram, a model makes the whole process transparent and open to scrutiny. This may reveal its weaknesses or confirm its validity—either way, it brings the methodology into the open. Second, a model stores, and thereby documents, your workflow, sparing you the task of keeping detailed external notes on what you did and why you did it. Third, models are flexible. They easily accommodate the kinds of changes to inputs or parameters mentioned above with little effort on your part. Fourth, models can be shared with colleagues, saving development work and encouraging collaboration on sound methods and best practices. Fifth, models run pretty fast—just wait and see.

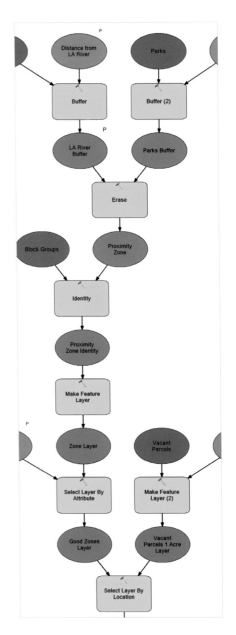

Lesson Seven roadmap

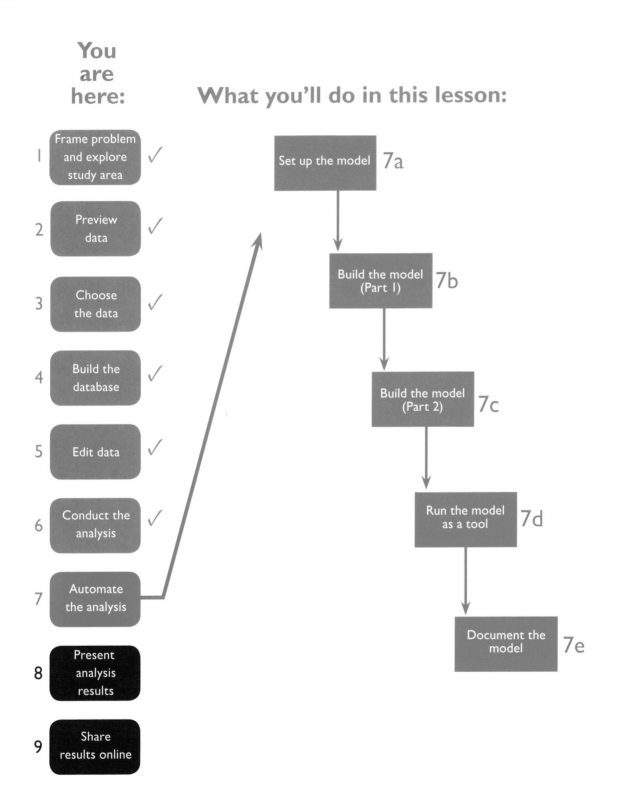

You are here:

1 Frame problem and explore study area ✓

2 Preview data ✓

3 Choose the data ✓

4 Build the database ✓

5 Edit data ✓

6 Conduct the analysis ✓

7 Automate the analysis

8 Present analysis results

9 Share results online

What you'll do in this lesson:

Set up the model 7a

Build the model (Part 1) 7b

Build the model (Part 2) 7c

Run the model as a tool 7d

Document the model 7e

Exercise 7a: Set up the model

ModelBuilder can be opened as an application from within either ArcMap or ArcCatalog. We'll use ArcMap for the same reasons we ran geoprocessing tools from ArcMap in Lesson 6: it's easy to add map layers as tool inputs and it's useful to be able to look at outputs in the map.

1) Create an output geodatabase. As in Lesson 6, we'll create a new geodatabase to hold our output data.

Ⓐ Start ArcMap and open a new blank document.

Ⓑ Open the Catalog window and pin it down to keep it open.

▷ If the Catalog tab isn't showing, click the Catalog window button 🗂 on the Standard toolbar.

Ⓒ Expand Folder Connections and navigate to C:\UGIS\ParkSite.

Ⓓ Right-click the AnalysisData folder and choose New → File Geodatabase.

Ⓔ Rename the new geodatabase **ModelBuilderOutputs**.

Ⓕ Right-click the ModelBuilderOutputs geodatabase and choose Make Default Geodatabase (Figure 7-1).

Figure 7-1

Ⓖ Save the map document in the MapsAndMore folder and name it **Lesson7**.

Understanding a model

To get the basic idea, let's forget about the LA River for a moment and prepare a Saturday morning breakfast. At right is a recipe (workflow) for making a dozen classic American pancakes from scratch. You should pick up the ingredients at your local store and make a batch to celebrate when you have completed this workbook!

The ingredients are symbolized by blue ellipses (inputs). These are prepared in various ways as shown by the orange rectangles (tools or processes). The results are green ellipses (outputs). You stir the basic ingredients to make the flour mix before adding the whipped egg, oil, and buttermilk. The mix should be properly blended before liquids are added; otherwise, you may get clumps of baking powder in the dough, which will lead to some very flat pancakes. After you make the dough, refrigerate it for several hours (or, even better, overnight) to allow time for chemical reaction. Then fry the dough in a pan with your favorite fat and eat it with a good brand of maple syrup. Delicious!

Can anything go wrong? If you follow the arrows, you're on a tried and true path. If you start improvising, you might make the recipe even better—but you might also get your weekend off to an indigestible start.

It's much the same with an ArcGIS workflow. With a complex analysis, it usually takes trial and error to get the recipe right. Typically, you'll experiment with several combinations of data, tools, and sequences before finding the best way to solve your problem. Once you find it, you want to save it and share it with friends, just like the pancake recipe.

We experimented a lot with the data and many ArcGIS tools before writing this cookbook (sorry, workbook). Having figured out the workflow, we implemented it in Lesson 6. But we also wanted to preserve it, encapsulate it, make it a formula—as you'll see, that's what ModelBuilder is all about.

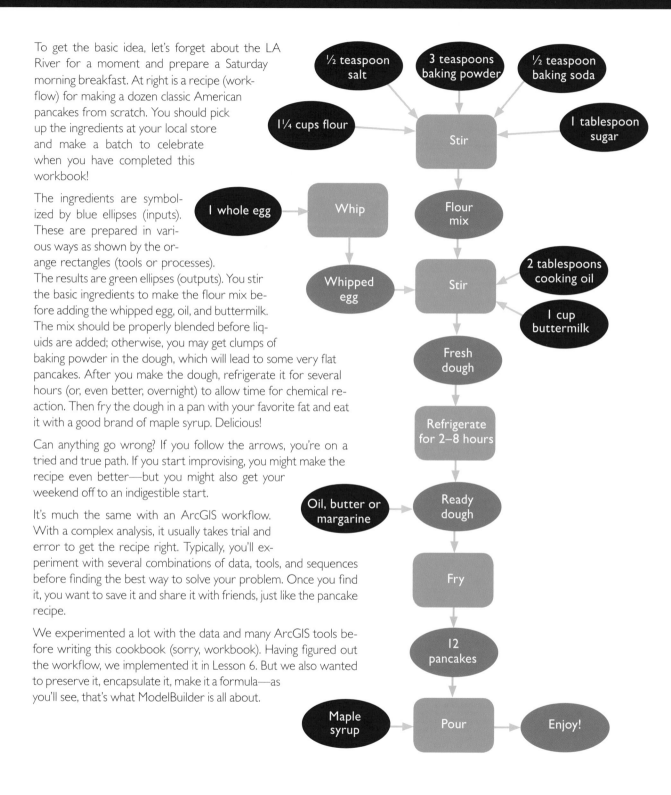

2) **Create a toolbox and model.** A model is a type of tool, and, as such, needs to go in a toolbox. It can't go in a system toolbox, however. We'll create a new toolbox in the MapsAndMore folder and put it there. We could also put it in My Toolboxes (under Toolboxes, at the root level of the Catalog window) or in any other folder or geodatabase.

Ⓐ At the top of the Catalog window, right-click the Home folder (ParkSite\MapsAndMore) and choose New → Toolbox.

Ⓑ Rename the new toolbox **ParkPlanningModels**.

Ⓒ Right-click the ParkPlanningModels toolbox and choose New → Model.

An empty Model window opens (Figure 7-2). This window can be resized and moved outside the ArcMap application window.

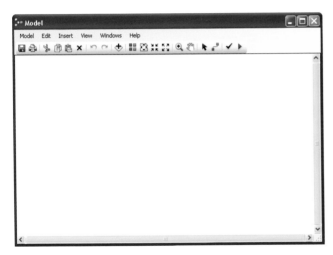

Figure 7-2

Ⓓ In the Catalog window, expand ParkPlanningModels.tbx, if necessary.

The model exists as a tool in the toolbox (Figure 7-3).

3) **Set model properties.** We'll rename the model and supply some information about it.

Ⓐ In the Model window, from the main menu, choose Model → Model Properties.

Ⓑ In the Model Properties dialog box, click the General tab.

Ⓒ Replace the default name with **ParkSuitabilityAnalysis** (no spaces).

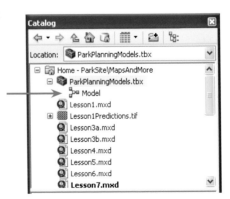

Figure 7-3

D Replace the default label with **Park Suitability Analysis** (this time include spaces).

The label is like a field name alias: a name that's more descriptive or easier to read.

E For the Description, enter: **This model identif ies vacant parcels near the Los Angeles River that may be suitable for a new park.**

F Optionally, add other information that will help identify the model, such as your name, your organization, and the date.

G Check the box to store relative path names.

The model, its input data, and its output data can now be moved without disruption as long as they remain in the same workspaces relative to each other.

H Make sure the "Always run in foreground" box is checked.

Ordinarily, geoprocessing tools run in the background so that you can keep using ArcMap while they work. Running the model in the foreground means that you have to stop and wait while the model executes. That's normally advisable because models are computationally intensive.

I Check that your entries match Figure 7-4, then click OK.

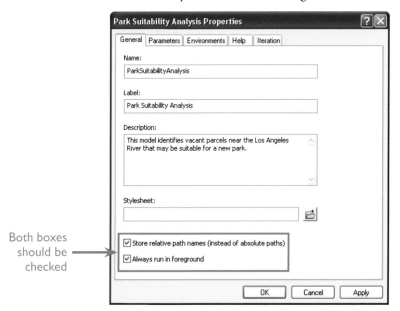

Both boxes should be checked

Figure 7-4

J On the ModelBuilder toolbar, at the top of the Model window, click the Save button 💾 to save the model.

In the Catalog window, the model now appears with its label.

K Optionally, unpin the Catalog window in ArcMap.

Exercise 7b: Build the model (Part 1)

Our model will re-create the analysis workflow from Lesson 6. Knowing how much work that entailed should give you insight into why models are such an efficient tool: not only for analysis projects, but for any series of operations that you might want to save, repeat, vary, or use as a basis for further development. Re-creating the analysis will let you focus your attention on the modeling process itself, since the data and geoprocessing tools will be familiar.

The data preparation tasks from Lesson 4 won't be part of the model. A big advantage of models is that it's easy to rerun them. We don't want to incorporate operations like adding fields, calculating field values, joining tables, and so on, that only need to be done once.

We're also going to leave out the step in Lesson 6 in which three nearby parcels were aggregated. That step was based on a human judgment that the gaps between the parcels were easements (unbuilt roads) and that it was legitimate to consolidate the parcels. We can't build that sort of judgment into the model.

In this exercise, we'll model about half of the analysis steps, saving the remainder for the next exercise. The stopping point is more or less arbitrary, but building a model incrementally is a good practice. It's easier to find and fix problems if you do test runs along the way.

1)   Add data to ModelBuilder. In this step, we'll add all the input data for the model. This is the opposite of what we did in Lesson 6, where we added layers one at a time as we needed them. Either way is fine in either ArcMap or ModelBuilder: it's a matter of preference.

Ⓐ On the ModelBuilder toolbar, click the Add Data or Tool button ✛.

▷ If you closed the model: in the Catalog window, right-click *Park Suitability Analysis* and choose Edit.

▷ If you closed ArcMap: start ArcMap and open the document *Lesson7.mxd,* then edit the model as described above.

Ⓑ In the Add Data or Tool dialog box, navigate to C:\UGIS\ ParkSite\AnalysisData\ReadyData.gdb.

Ⓒ Hold down the Control key and click these five feature classes (Figure 7-5):

- *BlockCentroids*
- *BlockGroups*
- *LARiver*
- *Parks*
- *VacantParcels*

Ⓓ Click Add.

Figure 7-5

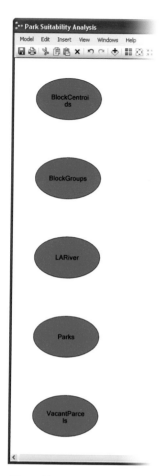

Figure 7-6

Each dataset is added as a blue ellipse to the Model window. (These blue ellipses are "input data variables," but we'll mostly just call them inputs.) The elements aren't automatically sized to fit; you may only be able to see a couple of them.

E Enlarge the Model window to about three-fourths of your screen space.

F On the ModelBuilder toolbar, click the Full Extent button ⬚.

You should see all five elements centered in the Model window and marked with blue selection handles.

The display of text within ellipses changes as you zoom in or out in the Model window. The places where words divide may change and so may the number of lines of text. Don't worry if the way text is displayed on your screen is different from the figures.

G Move the mouse pointer over any selected element, then click and drag the whole group to the left side of the Model window.

H Click in some empty white space in the Model window to unselect the elements (Figure 7-6).

2) **Add a tool to ModelBuilder.** The first step of the analysis in Lesson 6 was to buffer the LA River. Let's start with that.

A On the ModelBuilder toolbar, click the Add Data or Tool button ✛.

B In the Add Data or Tool dialog box, click the Up One Level button ⬆ as many times as you can—all the way to the root level of the catalog.

C Double-click the Toolboxes folder, then System Toolboxes, then Analysis Tools, then the Proximity toolset.

D Click the Buffer tool to highlight it (Figure 7-7), then click Add.

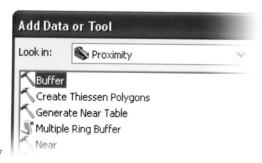

Figure 7-7

The tool is added to the Model window, connected by a small arrow to an Output Feature Class (Figure 7-8). The tool and its output are white because they're not ready to run. The tool needs to be connected to input data and its parameters have to be filled out.

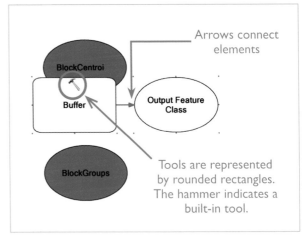

E Make sure the Buffer tool and Output Feature Class elements are selected (marked with blue selection handles).

▹ If they're not, hold down the Shift key and click to select them. You don't have to select the arrow.

F Drag the selected elements down and place them to the right of the *LARiver* input data element. (The exact position doesn't matter.)

G On the ModelBuilder toolbar, click the Connect tool 📟 and move the mouse pointer over the blue *LARiver* ellipse.

Figure 7-8

The cursor looks like a magic wand ※﹨.

H From the *LARiver* ellipse, drag and drop a blue arrow on top of the Buffer tool (Figure 7-9), then release the mouse button.

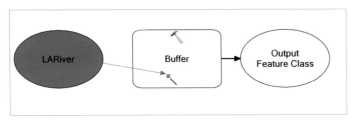

Figure 7-9

You are prompted for the type of connection to make (Figure 7-10).

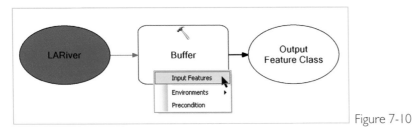

Figure 7-10

I Click Input Features.

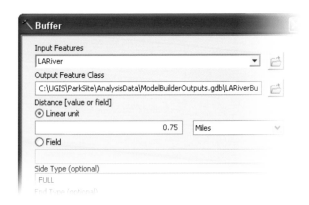

Figure 7-11

J Right-click the Buffer tool and choose Open.

The familiar geoprocessing tool opens. The input features are set to *LARiver,* as specified by the connection arrow, and the output feature class is directed to the default geodatabase. We'll rename this feature class in accordance with our convention.

K Highlight and delete the existing output feature class path. Type **LARiverBuffer** and press Tab.

  ▷ Or just remove the underscore from "LARiver_Buffer."

L For the linear unit, type **0.75**. Set the adjacent drop-down arrow to Miles (Figure 7-11). Click OK.

In ordinary geoprocessing, clicking OK would run the process; in ModelBuilder it *prepares* the process to run. The tool and output data elements turn yellow and green, respectively, which means they are properly connected and filled out (Figure 7-12).

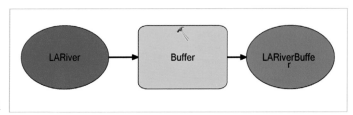

Figure 7-12

We now have a process ready to run—a very simple model—but we're not going to run it yet.

The green element displays the name of the output feature class. In the figure above, the line breaks badly, stranding the letter "r" on a line by itself. This effect is a function of many things: the zoom level, the text length, the font, and the size of the ellipse. It probably looks the same on your screen, but even if it doesn't, you'll see the problem elsewhere. We'll fix the problem by changing the text. (One alternative is to resize the ellipse, and another is just to ignore it.)

M Right-click the *LARiverBuffer* element. On the context menu, choose Rename.

N In the Rename box, replace the existing name with **LA River Buffer** (Figure 7-13). Click OK.

Figure 7-13

The text is reformatted. Word spaces help the line break naturally (Figure 7-14).

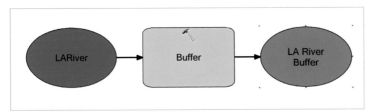

Figure 7-14

The name is just for display. It will be given to layers added to ArcMap from the model, but it won't change the output feature class name.

- ⊙ Right-click the green *LA River Buffer* output and choose Open (Figure 7-15).

Figure 7-15

display name ——→

output feature class

The display name has spaces but the output feature class name doesn't.

- ⊙ Close the LA River Buffer dialog box.
- ⊙ On the ModelBuilder toolbar, click the Save button 🖫 .

3) Buffer the parks. In Lesson 6, after buffering the river, we created ¼-mile buffers around existing parks.

- Ⓐ On the ModelBuilder toolbar, click the Add Data or Tool button ✛.

The Add Data or Tool dialog box should open to the Proximity toolset, and the Buffer tool should be right there.

- Ⓑ Click the Buffer tool to highlight it, then click Add.

As before, the tool and its output feature class are added to the model. The tool is named Buffer (2) to distinguish it from the other Buffer tool.

- Ⓒ On the ModelBuilder toolbar, click the Select tool ➤ .
- Ⓓ Drag the selected elements down and place them to the right of the *Parks* input data element.
  - ▷ If you don't click the Select tool, and your cursor is still a magic wand, you'll draw a connector. Press the Escape key to delete it.

Figure 7-16

Last time, we set the input features in the Buffer tool by drawing a connector. Now, we'll do it another way.

ⓔ Right-click the Buffer (2) tool and choose Open.

▷ Or double-click the tool to open it.

ⓕ Click the Input Features drop-down arrow and choose *Parks*.

The available inputs are marked with a blue recycle symbol (for "data variables") in contrast to the familiar symbol for layers—there actually aren't any layers in the map document. Every input data element is technically a variable because you can open it and point it to another feature class.

ⓖ For the linear unit, type **0.25**. Set the units drop-down arrow to Miles.

ⓗ Click the Dissolve Type drop-down arrow and choose ALL.

We don't have the visual reason to dissolve the buffers that we had in Lesson 6 (we're not going to look at this tool's output), but it doesn't cause us any trouble to duplicate the workflow.

ⓘ Compare your dialog box to Figure 7-16, then click OK.

Both buffer processes should now be colored (Figure 7-17).

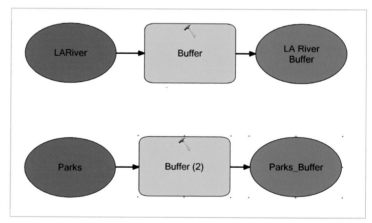

Figure 7-17

4)   **Erase park buffers.** Now, as in Lesson 6, we'll use the Erase tool to eat away the park buffers from within the LA River buffer.

ⓐ Position the ModelBuilder window so you can see the right side of the ArcMap application window.

▷ If necessary, make the Model window smaller or slide it partway off the computer screen.

B In ArcMap, open the Search window.

▷ If the Search tab isn't there, click the Search window button 🔍 on the Standard toolbar in ArcMap.

C Near the top of the Search window, click Tools to limit the search to tools.

D In the Search box, type **erase** and press Enter.

▷ Or pick "erase (analysis)" from the drop-down list.

The Erase (analysis) tool is returned at the top of the list of tools.

E Place the mouse pointer over the tool name, as in Figure 7-18.

Figure 7-18

F Drag and drop the tool into the Model window to the right of the output data elements (Figure 7-19).

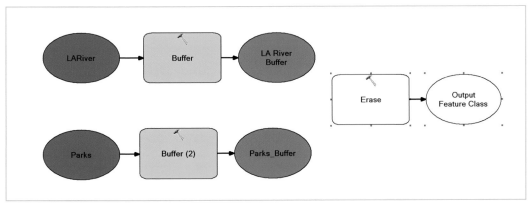

Figure 7-19

G Double-click the Erase tool to open it.

H Set the input features to *LA River Buffer* and the erase features to *Parks_Buffer*.

❶ Compare your Erase dialog box to Figure 7-20, then click OK.

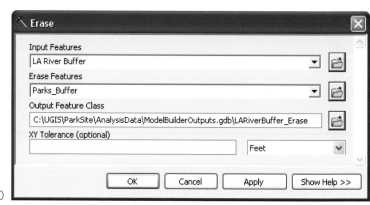

Figure 7-20

In the model, the output datasets of the Buffer tools are now connected as inputs to the Erase tool.

The name of the Erase tool output is long. We don't care about the feature class name, because we don't plan to save it, but we want the name to look better in the model.

❶ Right-click the *LARiverBuffer_Erase* element and choose Rename.

❶ Rename the element **Proximity Zone** (this is similar to its name in Lesson 6) and click OK (Figure 7-21).

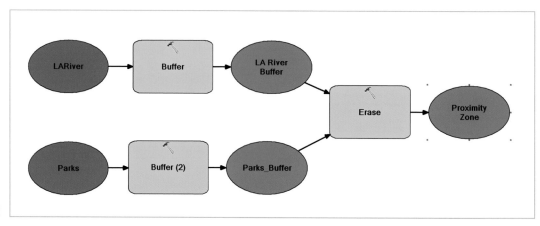

Figure 7-21

ⓛ Right-click the *Proximity Zone* element again to open its context menu.

Notice that the Intermediate setting is grayed-out (Figure 7-22).

ⓜ Click in some empty white space to close the context menu and unselect the element.

ⓝ Right-click any blue input data element to see that its Intermediate setting is grayed-out, too.

ⓞ Right-click the *Parks_Buffer* element and check its status.

A check mark flags this element as Intermediate.

ⓟ Right-click the *LA River Buffer* element.

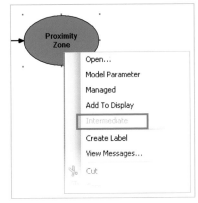

Figure 7-22

It's also flagged as Intermediate. Since this dataset is one we want to keep in the output geodatabase, we'll unflag it.

ⓠ On the context menu, click Intermediate to uncheck it.

▷ Optionally, open the context menu again to confirm that the check mark is gone.

Intermediate data

Every data element in ModelBuilder is either intermediate or not. Intermediate data is data that ModelBuilder assumes you don't ultimately want to keep. Input data elements (blue ellipses) can't be intermediate. Output data elements (green ellipses) that come at the end of a chain can't be intermediate either. All other output data elements are flagged as intermediate. When you run a model as a tool (Exercise 7D), intermediate data is deleted automatically and never appears in the output geodatabase. When you run a model from the Model window, intermediate data is still written to the output geodatabase, but can subsequently be deleted with the Delete Intermediate Data command on the Model menu. Any data element that is flagged as intermediate can be unflagged if you want to save it.

5) Assign block group demographics to the proximity zone. After erasing the park buffers in Lesson 6, we used the Identity tool to overlay the remaining area of interest with block group attributes (Exercise 6b). We'll do the same thing now in the model.

ⓐ In the ArcMap Search window, delete the existing search term, type **identity**, and press Enter.

ⓑ From the list of tools, drag and drop Identity (Analysis) into the Model window near the Proximity Zone output element.

ⓒ Open the tool. Set the Input Features to *Proximity Zone* and the Identity Features to *BlockGroups*.

D Compare your Identity tool dialog box to Figure 7-23, then click OK.

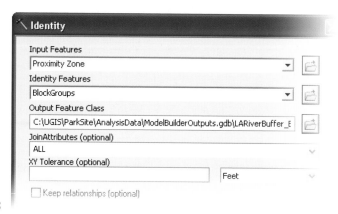

Figure 7-23

The Identity tool is colored in and connected to its two inputs. At this point, the model layout is getting complicated. We'll adjust it after we rename the output element of the Identity tool.

E Rename LARiverBuffer_Erase_Identity to **Proximity Zone Identity**. Click OK.

F Unselect the output data element by clicking in white space.

G On the ModelBuilder toolbar, click the Auto Layout button ▊▊, then click the Full Extent button ❖.

Your layout should look pretty much like Figure 7-24.

H Save the model.

Figure 7-24

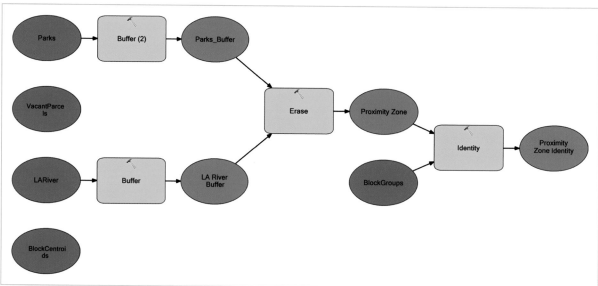

6) Validate and run the model. This is a good point to test-run the model. Before running a model, you should validate it to make sure that all is well with the data variables and tool parameters.

ModelBuilder will write the output data to the ModelBuilderOutputs geodatabase. By default, it doesn't add layers to ArcMap. If we want to see the data, we have to tell ModelBuilder to add it (or add it from the geodatabase later).

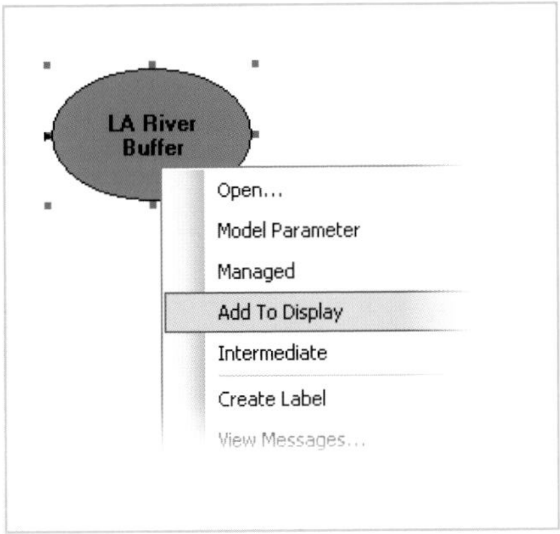

Ⓐ In the Model window, right-click the *LA River Buffer* output data element and choose Add To Display (Figure 7-25).

Ⓑ In the same way, add *Proximity Zone Identity* to the display.

Ⓒ On the ModelBuilder toolbar, click the Validate Entire Model button ✔.

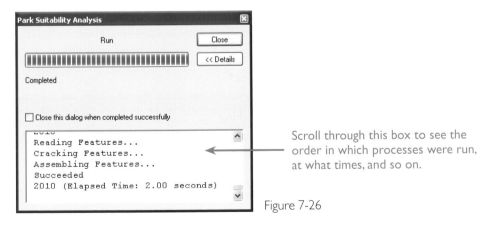

Figure 7-25

Validating models

Validation ensures that a model has access to its source data. If you rename, move, or delete a feature class referenced by an input data variable, that variable and its dependent processes will turn white (not ready) when you validate the model. Similarly, if you delete a field from a table, and that field is specified by a tool's parameter, the tool and its dependent processes will turn white when you validate.

The next step is to run the model. As it runs, you should get a visual progress report, with each tool turning red as it is processed. (If you don't see the red highlight, it's okay—the job is still being done.)

Ⓓ On the ModelBuilder toolbar, click the Run button ▶.

When the model finishes running, you get a message box (Figure 7-26).

Figure 7-26

Scroll through this box to see the order in which processes were run, at what times, and so on.

Ⓔ Close the message box.

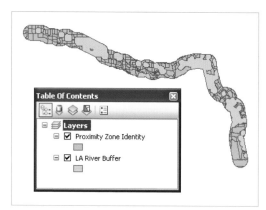

Figure 7-27

Figure 7-28

In the Model window, the tools and output data elements have drop shadows. This tells you that the model has been run.

F Minimize the Model window or move it away from the ArcMap window.

The two outputs marked Add To Display should appear as layers in ArcMap (Figure 7-27). If they don't show up, you can add them from the ModelBuilderOutputs geodatabase. The feature class names are *LARiverBuffer* and *LARiverBuffer_Erase_Identity*.

G In the Catalog window, under AnalysisData, expand the ModelBuilderOutputs geodatabase.

It contains four feature classes (Figure 7-28). They have the names assigned by the tool parameters in ModelBuilder. The ArcMap layers have the display names given to the output data elements.

H In the Table of Contents, open the attribute table for the *Proximity Zone Identity* layer. Scroll across to see its attributes.

The table has 298 records and includes the demographic attributes from the *BlockGroups* feature class. It's the same result that we got in Lesson 6 on page 205.

I Close the attribute table.

J In the Table of Contents, remove the *Proximity Zone Identity* and *LA River Buffer* layers from the map.

K Restore the Model window, or move it back into the center of your screen.

L Save the model.

We'll finish the model in the next exercise. If you want, you can take a break here.

M Optionally, from the ModelBuilder main menu, choose Model → Close.

N Optionally, exit ArcMap. If you exit, don't save your changes.

Adding layers to the display

After the model has been run, you can right-click any output data element in the Model window and choose Add To Display to add it as a layer to ArcMap.

Exercise 7c: Build the model (Part 2)

In this exercise, we'll keep following the Lesson 6 workflow and finish the analysis model. As mentioned, we're not going to aggregate the three candidate parcels that are close to each other, so our results won't be identical to Lesson 6.

In this exercise, we'll see how ModelBuilder handles operations on layers. In Lesson 6, we used attribute and spatial queries to narrow down the number of potential park sites. Queries (and, for that matter, any type of feature or record selections) are not made directly on feature classes, but rather on layers. To make feature selections in ModelBuilder, we therefore need a special tool that turns an input feature class into a layer. This tool is called Make Feature Layer. When we make layers in ModelBuilder, it's usually for the purpose of making a selection that can be used as an input to further processes. For a specialized use of the Make Feature Layer tool in ArcMap, see *Apportioning attribute values* on page 223.

1) Select features by demographic attributes. In Lesson 6, we used an attribute query to select features in the proximity zone according to demographic values. We'll re-create that query now.

Ⓐ If necessary, start ArcMap. In the Getting Started dialog box, open Lesson7 from the list of recent maps.

▷ If ArcMap and ModelBuilder are still open from the previous exercise, skip to Step 1C.

Ⓑ In the Catalog window, in the Home folder, expand the ParkPlanningModels toolbox. Right-click *Park Suitability Analysis* and choose Edit.

Ⓒ Move the Model window far enough out of the way that you can use the ArcMap Search window.

Ⓓ In the ArcMap Search window, click Tools, if necessary, then search for `select by attribute`.

In the list of returned tools, Select Layer By Attribute (Data Management) is the one we want (Figure 7-29). The tool's name tells us that it needs to work on a layer.

Ⓔ Drag and drop Select Layer By Attribute into the Model window. Don't worry about where it goes exactly.

Ⓕ Double-click the tool to open it.

Figure 7-29

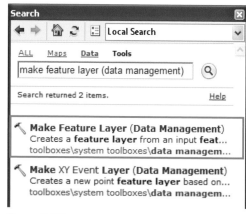

Figure 7-30

In the tool dialog box, there's no drop-down list for the required input (Layer Name or Table View) because there aren't any layers or tables in the ArcMap document.

Ⓖ Close the Select Layer By Attribute dialog box.

Ⓗ In the ArcMap Search window, search for **make feature layer**.

In the list of returned tools, we want Make Feature Layer (Data Management) (Figure 7-30).

Ⓘ Drag and drop Make Feature Layer into the Model window, then open the tool.

Ⓙ Set the Input Features to *Proximity Zone Identity*.

Ⓚ Replace the Output Layer path with **Zone Layer** (Figure 7-31), then click OK.

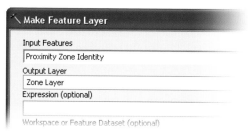

Figure 7-31

The tool is connected to the *Proximity Zone Identity* element and filled in.

The layout may be different, but the elements in your Model window should be connected as in Figure 7-32.

Ⓛ Right-click the *Proximity Zone Identity* element.

This dataset is now checked Intermediate because it has become an input to a tool. We're not going to look at it again on the map, so we'll unflag its display property.

Ⓜ Uncheck Add To Display.

Ⓝ Open the Select Layer By Attribute tool.

Ⓞ Set the Layer Name or Table View to *Zone Layer*.

This choice is available now because the Make Feature Layer tool has defined its existence.

Ⓟ Next to the Expression box, click the SQL button 🔲 to open the Query Builder.

Figure 7-32

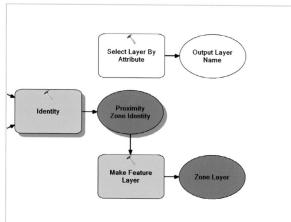

Q Build the expression shown in Figure 7-33:

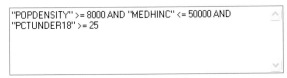

"POPDENSITY" >= 8000 AND "MEDHINC" <= 50000 AND
"PCTUNDER18" >= 25

Figure 7-33

R Click OK on the Query Builder dialog box. Confirm that the Select Layer By Attribute dialog box looks like Figure 7-34, then click OK.

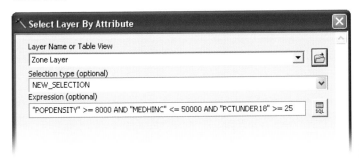

Figure 7-34

S Rename the output data element from Zone Layer (2) to **Good Zones Layer**.

T On the ModelBuilder toolbar, click the Auto Layout button ▪▪, then the Full Extent button ▩.

The model may be too big now for you to make out the names of individual elements when you're zoomed to the full extent.

U On the ModelBuilder toolbar, click the Zoom In tool ⊕ and zoom in on the right side of the model.

V From the ModelBuilder main menu, choose View → 75% (Figure 7-35).

Figure 7-35

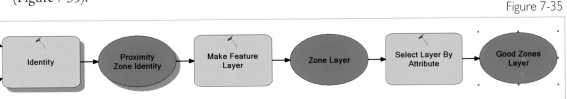

W Save the model.

Zooming and panning your model

The ModelBuilder toolbar has several navigation tools for zooming in and out on your model: ▩ ▪▪ ▪▪ ⊕ and ✋. You can also use the horizontal and vertical scroll bars. The View menu includes some of the same tools as well as predefined and custom zoom level choices.

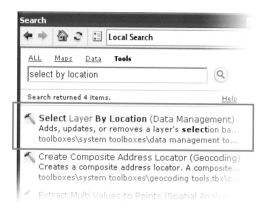

Figure 7-36

2) Select parcels within good zones. The attribute query for desirable demographics significantly narrows down the "good zones" where we can look for vacant parcels. We now have the ModelBuilder equivalent of an attribute selection on a layer, bringing us to the point in the analysis where we were at the end of Lesson 6b. The next thing to do is to make a spatial query for vacant parcels that lie within the good zones.

Ⓐ In the ArcMap Search window, search for **select by location**.

In the list of returned tools, the one we want is Select Layer By Location (Data Management) (Figure 7-36).

Ⓑ Drag and drop Select Layer By Location into the Model window.

Since this tool also requires a layer as input, we need to make a feature layer from the *VacantParcels* input element.

Ⓒ In the ArcMap Search window, search for **make feature layer**. Drag and drop the Make Feature Layer tool into the model.

Ⓓ Open the Make Feature Layer (2) tool and set the Input Features to *VacantParcels*.

In Lesson 6 we did a spatial query for vacant parcels in good zones, followed by an attribute query for parcels one acre or larger. We can shortcut the attribute query on parcel size by incorporating the query in the Expression parameter of the Make Feature Layer tool. (We actually could have done this in the previous step, too.)

Building the query in the Expression parameter of the Make Feature Layer tool switches the order of the two query operations from Lesson 6, but doesn't affect the logic. It doesn't matter whether we first select the vacant parcels in good zones and then subselect the one-acre ones (Lesson 6) or first select all the one-acre vacant parcels and then subselect the ones in good zones (as we'll do here).

Ⓔ Next to the Expression box, click the SQL button.

Ⓕ In the Query Builder, create the expression shown in Figure 7-37, then click OK.

Figure 7-37

Ⓖ Replace the default Output Layer name with **Vacant Parcels 1 Acre Layer**.

ⓗ Confirm that the dialog box looks like Figure 7-38, then click OK.

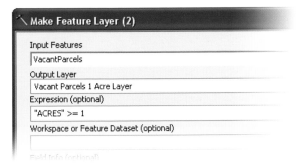

Figure 7-38

ⓘ Click Auto Layout, then zoom to the full extent to see how the model has been rearranged.

ⓙ Zoom in to a more legible level, like 75% or 100%.

ⓚ Optionally, rename any elements with bad line breaks by inserting spaces in their names.

  ▷ For example, rename the *VacantParcels* input data element to **Vacant Parcels**.

The part of the model that you're working with should look like Figure 7-39.

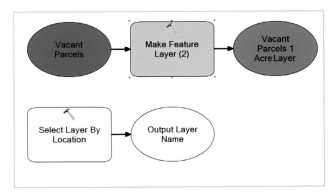

Figure 7-39

ⓛ Open the Select Layer By Location tool and set the Input Feature Layer to *Vacant Parcels 1 Acre Layer*.

ⓜ If necessary, set the Relationship to INTERSECT.

ⓝ Set the Selecting Features to *Good Zones Layer*.

Vacant one-acre parcels will be selected if they intersect good zones.

O Make sure your Select Layer By Location dialog box looks like Figure 7-40, then click OK.

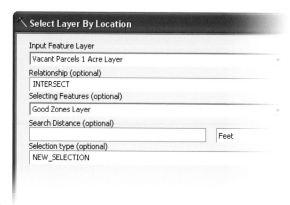

Figure 7-40

P Click the Auto Layout button and scroll to the right side of the Model window.

The name *Vacant Parcels 1 Acre Layer* is reused for the output because no new data is created. All that's happening is that a new selection is being made on the layer. But it might be clearer if we give the new element its own name.

Q Rename the last output data element to **Good Zone Parcels Layer** (Figure 7-41).

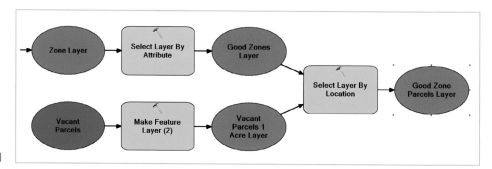

Figure 7-41

3) **Copy features.** In Lesson 6, we used the Export Data command to write our selection to a feature class, but in ModelBuilder we have to rely on geoprocessing tools.

A In the ArcMap Search window, locate the Copy Features (Data Management) tool. Drag and drop it into the Model window.

B Open the tool and set the Input Features to *Good Zone Parcels Layer.*

C Replace the Output Feature Class path with **RecommendedSites** and press Tab.

ⓓ Make sure the Copy Features dialog box looks like Figure 7-42, then click OK.

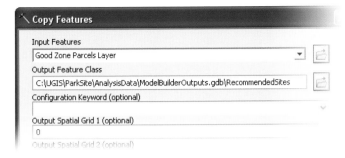

Figure 7-42

ⓔ Click the Auto Layout button and the Full Extent button.

ⓕ Save the model.

At this point in Lesson 6, we aggregated the three sites that were right by each other. Since we're not doing that here, we'll move on to create access zones around the candidates and count the total population inside them.

4) Create park access zones. The access zones are quarter-mile buffers. Rather than add the Buffer tool again from the Search window, we'll copy and paste it from inside the model.

ⓐ Zoom in on the left side of the model.

ⓑ Right-click the Buffer (2) tool and choose Copy.

ⓒ Right-click in some empty white space in the Model window and choose Paste.

ⓓ Click Auto Layout (Figure 7-43).

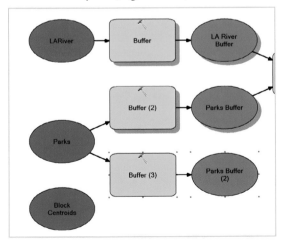

Figure 7-43

A Buffer (3) tool is created. Like Buffer (2), it's connected to the *Parks* input data. All the parameters of the Buffer (2) tool have been copied as well.

E Open the Buffer (3) tool and set the Input Features to *RecommendedSites*.

The yellow icon ⚠ by the Output Feature Class parameter warns you that another feature class already has the name *Parks_Buffer*.

F Replace the Output Feature Class path with **ParkAccessZones** and press Tab.

G Keep the same distance value (0.25 miles) but change the Dissolve Type to NONE.

H Make sure the Buffer (3) dialog box looks like Figure 7-44, then click OK.

Figure 7-44

I Click Auto Layout and scroll to the right side of the Model window.

J Rename the output of the Buffer (3) tool to **Park Access Zones**.

5) Count total population within park access zones. To get a population count, we'll spatially join *BlockCentroids* to *ParkAccessZones*. The output will have the same features as *ParkAccessZones* and the attributes of both inputs. A merge rule will sum the populations of the block centroids within each access zone.

A In the ArcMap Search window, search for **spatial join**.

B Drag and drop the Spatial Join (Analysis) tool into the Model window.

C Open the tool and set the Target Features to *Park Access Zones*.

D Set the Join Features to *Block Centroids*.

ⓔ Accept the default Output Feature Class path.

We'll delete unnecessary fields from the output attribute table. Delete carefully so that your mouse clicks don't get ahead of you.

ⓕ In the Field Map of Join Features area, click the first field in the list (it should be SHAPE_LENGTH) to highlight it.

ⓖ Click the Delete button ✖ to delete the field.

ⓗ Go on to delete all the fields except ACRES (now second from the top) and POP2000 (third from the bottom).

▷ If you accidentally delete either of these two fields, close the tool and start again.

ⓘ Right-click the POP2000 field and choose Merge Rule → Sum.

Underneath the Field Map of Join Features box is the Match Option parameter, which defines the spatial relationship to be evaluated. Either INTERSECT (which includes containment) or CONTAINS will work.

ⓙ Accept the default Match Option setting of INTERSECT.

ⓚ Compare your dialog box to Figure 7-45, then click OK.

ⓛ Rename the output of the Spatial Join tool to **Park Access Zones Pop**.

ⓜ Do an auto layout, then zoom to the full extent.

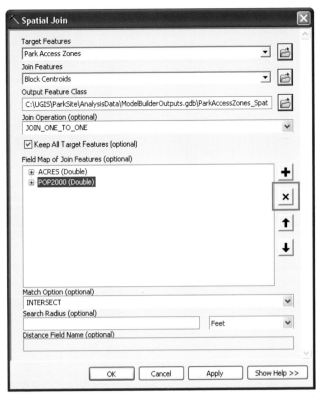

Figure 7-45

The model is essentially done. We'll skip the distance calculation to the river, as well as the overlays that brought the demographic attributes into *RecommendedSites* in Lesson 6. The last thing we'll do before running the model is set the status of our output data elements correctly. The three outputs that we want to save to the geodatabase and add as layers to ArcMap are:

- *LA River Buffer*
- *Recommended Sites*
- *Park Access Zones Pop*

All other outputs are intermediate data.

ⓝ Right-click each green output data element and confirm or set its Add To Display and Intermediate status as shown in Figure 7-46.

Figure 7-46

Output Element	Add To Display	Intermediate
Parks Buffer		√
LA River Buffer	√	
Proximity Zone		√
Proximity Zone Identity		√
Zone Layer		√
Vacant Parcels 1 Acre Layer		√
Good Zones Layer		grayed-out ←
Good Zone Parcels Layer		grayed-out ←
Recommended Sites	√	
Park Access Zones		√
Park Access Zones Pop	√	grayed-out

Grayed-out because they are selections on a layer

O Save the model and zoom to the full extent.

6) **Validate and run the model.** We're ready to run the model.

A On the ModelBuilder toolbar, click the Validate Entire Model button ✔.

This removes drop shadows from elements that have already been run and returns them to their ready-to-run state.

B Click the Run button ▶.

C When the model finishes, close the dialog box. Minimize the Model window or move it out of the way.

7) **View results.** If the model ran successfully, you should see the *Park Access Zones Pop, Recommended Sites,* and *LA River Buffer* layers in ArcMap.

▷ If you don't see these layers, restore the Model window. Check or uncheck Add To Display for output data elements as needed.

A Drag *Recommended Sites* to the top of the Table of Contents.

▷ If necessary, click the List By Drawing Order button at the top of the Table of Contents.

B Optionally, resymbolize the layers.

C Open the *Park Access Zones Pop* attribute table (Figure 7-47).

Figure 7-47

Shape *	Join_Count	TARGET_FID	ACRES	POP2000	Shape_Length	Shape_Area
▶ Polygon	29	1	1.494599	3500	9560.14804	7212417.05940
Polygon	30	2	2.362142	3535	9872.479575	7675524.29237
Polygon	27	3	1.243729	3011	9302.453707	6859361.22658
Polygon	25	4	1.658459	1210	9447.834421	7077155.15383
Polygon	15	5	3.433673	1262	9963.8075	7838749.71726
Polygon	14	6	1.950277	2956	9599.382878	7297054.66129
Polygon	29	7	1.711665	3309	9720.528802	7438949.64177

The table has seven records instead of five because of the three sites that weren't aggregated. The values in the ACRES and POP2000 fields should match your Lesson 6 results for four of the records.

D Close the table. Optionally, zoom in for a look at any of the suitable sites.

E Remove all layers from the Table of Contents.

F In the Catalog window, expand the ModelBuilderOutputs geodatabase, if necessary.

Figure 7-48

It should contain seven feature classes. There are eleven output elements in the model, but four of them are layers.

ⓖ Restore the Model window. From the main menu, choose Model ➙ Delete Intermediate Data.

Data marked Intermediate in the model is deleted from the geodatabase. You should be left with the three feature classes shown in Figure 7-48.

ⓗ Save and close the model.

ⓘ Save the map document. Optionally, exit ArcMap.

Troubleshooting a model

If a model process fails or a warning is issued, you'll see it in the message box. Warnings appear with a code that links to a description in the ArcGIS Desktop Help. One way to check whether individual outputs are okay is to add them to the display so you can see the layer in ArcMap. If you combine this technique with incremental building and testing of the model, you can save yourself a lot of backtracking when things go wrong. If you add processes to a model and rerun it without validating the entire model, only the "downstream" part (the part without drop shadows) will be executed.

Exercise 7d: Run the model as a tool

In this exercise, we'll set up the model as a geoprocessing tool. This is an efficient way to run a model, and it's a good way to share the model with people who may want to use it but may not be very interested in the details of its construction. When you run a model as a tool, it has a dialog box interface like any other geoprocessing tool: the user sets a number of parameters and then clicks OK. Of course, behind the scenes, the model is a long chain of tools, each of which has its own cluster of required and optional parameters: if all these parameters had to be set by the user in the model's dialog box, it would be a long process.

Fortunately, that's not how it works. When you set up the model to run as a tool, you specify which parameters from which tools inside the model will be exposed in the model dialog box. Only these parameters are filled out by the user (you can supply default values): all others are preset by you, and the person running the tool never sees them. In other words, you decide how flexible to make the tool.

The ability to test sensitivity to changes in variables and to explore alternative outcomes are key reasons for building models in the first place. Accordingly, when we set up the model to run as a tool, we should think about which factors the user is most likely to want to experiment with.

1) **Open the model.** In previous exercises, you've *edited* the model, which is how you start work in the Model window. To run the model as a geoprocessing tool, you *open* it.

- Ⓐ If necessary, start ArcMap and open Lesson7 from the list of recent maps.

- Ⓑ In the Catalog window, in the Home folder, expand the ParkPlanningModels toolbox. Right-click *Park Suitability Analysis* and choose Open.

The model opens as a geoprocessing tool (Figure 7-49). It's just an empty gray dialog box with the message, "This tool has no parameters." That's because we haven't yet gone into the Model window and chosen the parameters we want to expose.

Figure 7-49

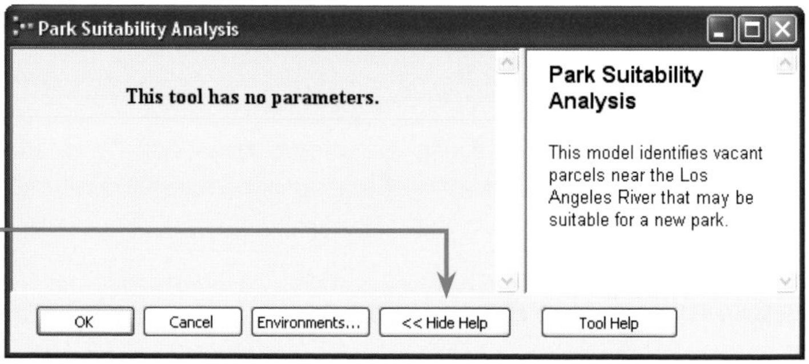

If the tool's help panel is showing, you'll see the text you entered in the Model Properties in Exercise 7A.

- Ⓒ Close the Park Suitability Analysis dialog box.
- Ⓓ In the Catalog window, right-click the model and choose Edit.
- Ⓔ In the Model window, zoom in on the left side of the model, then set your view scale to a comfortable level.
  - ▷ How? Click the View menu and choose a value.

2) **Make tool parameters for buffer size.** Users of the model tool will probably want to experiment with our buffer sizes for maximum distance from the river (0.75 miles) and minimum distance from existing parks (0.25 miles). To expose these parameters to the tool's dialog box, we first have to define them as variables. This allows their values to be changed any time the model is run. Then we flag the variables as "model parameters" (parameters we want to include in the model's tool dialog box).

- Ⓐ In the Model window, right-click the Buffer tool, connected to the *LARiver* input, and choose Make Variable → From Parameter → Distance [value or field].

This says that we want the distance parameter in the Buffer tool to be a variable. A small light blue ellipse, called a value variable, is added to the Model window.

❸ On the ModelBuilder toolbar, click the Select tool ⬉, if necessary. Click once on the value variable so that it remains selected
while the Buffer tool is unselected.

❹ Drag the value variable slightly away from the *LARiver* element.

❺ Right-click the variable and choose Open (Figure 7-50).

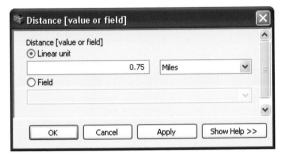

Figure 7-50

Within the Model window, the Distance value is now detached from
the Buffer tool and can only be set here. We'll use the current value
of 0.75 miles as the default for the model tool.

❺ Close the Distance dialog box.

❻ Right-click the value variable and choose Model Parameter.

The letter "P" (for "parameter") appears above the variable (Figure
7-51).

In the model dialog box, the title of the parameter will match the
variable name. We don't really want the parameter to be called
"Distance [value or field]."

❻ Right-click the value variable and choose Rename. For the new
name, type **Distance from LA River** (Figure 7-52), then
click OK.

Figure 7-52

We'll resize the element so we can read the text.

❻ With the value variable selected, click and drag one of its corner
selection handles until you can see the whole name (Figure
7-53).

Figure 7-51

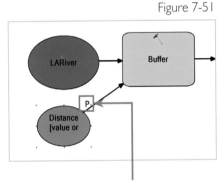

The variable is flagged as a
tool parameter.

Figure 7-53

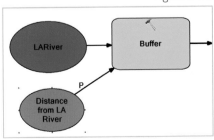

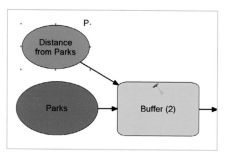

Figure 7-54

We'll make a parameter for the *Parks* buffer distance in the same way.

I Right-click the Buffer (2) tool, connected to the *Parks* input, and choose Make Variable → From Parameter → Distance [value or field].

J Click on the new variable so that it's the only selected element, then move it off the *Parks* input element.

K Right-click the variable and choose Model Parameter.

L Rename the variable **Distance from Parks** and resize it as needed (Figure 7-54).

3) **Make parameters for demographics and park size.** We'll make tool parameters from the query expressions for demographics and park size so we can experiment with these, too.

A Scroll to the middle of the model.

B Right-click the Select Layer By Attribute tool and choose Make Variable → From Parameter → Expression.

A value variable with the name Expression is added to the Model window.

C Make the new variable the only selected element, then move it to a good place.

D Hover the mouse over the variable element to see the contents of the expression as a tip (Figure 7-55).

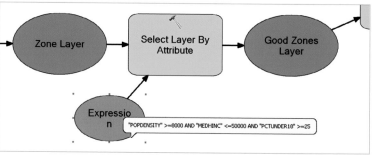

Figure 7-55

Figure 7-56

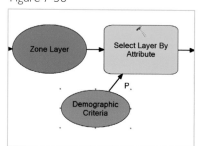

E Flag the variable as a model parameter.

F Rename the variable **Demographic Criteria** and resize it as needed (Figure 7-56).

▷ You can drag a side selection handle to widen the ellipse without changing its height.

G Right-click the Make Feature Layer (2) tool, connected to the *Vacant Parcels* input, and choose Make Variable → From Parameter → Expression.

H Make the new variable the only selected element, then move it to a good place.

I Hover the mouse over the element to see its expression ("ACRES" >= 1).

J Set the variable as a model parameter.

K Rename the variable **Park Size in Acres** and resize it as needed (Figure 7-57).

L Click the Validate Entire Model button ✔.

M Save and close the model.

4) Run the tool. Now we'll reopen the model as a tool and run it.

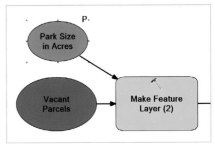

Figure 7-57

A In the Catalog window, right-click *Park Suitability Analysis* and choose Open.

The tool displays the parameters you've specified (Figure 7-58). We'll experiment with new values on the run after this one. This time, we'll run the model with default values to make sure it works.

B Click OK on the Park Suitability Analysis dialog box.

C When the model is finished running, close the message box.

The model gives the same results whether you run it from ModelBuilder or as a tool, but its execution is a little different. For one thing, when the model is run as a tool, intermediate data is deleted automatically.

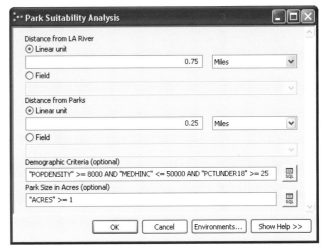

Figure 7-58

D In the Catalog window, expand, if necessary, the ModelBuilderOutputs geodatabase.

There should be three feature classes:

- *LARiverBuffer*
- *ParkAccessZones_SpatialJoin*
- *RecommendedSites*

By default, ArcMap doesn't allow geoprocessing outputs to overwrite each other, yet these three feature classes replaced three of the same name that were there before. Also by default, ArcMap *does* add geoprocessing results as layers to the map, but no layers were added here. As we'll see next, the model outputs have to be exposed as tool parameters for these geoprocessing defaults to take effect.

Reordering tool parameters

To change the order in which parameters are displayed in the dialog box, open the Model Properties and click the Parameters tab. The parameters appear in a list where they can be selected and moved up or down with arrow keys. The Model Properties can be opened from the Model menu in ModelBuilder, or from the model's context menu in the Catalog window.

5) Make tool parameters for the outputs. Flagging the outputs of the model as tool parameters will let us add the results as layers to ArcMap. More importantly, it will give us control over whether output datasets overwrite each other. As we experiment with different values in successive runs of the model, we may want to save our results with different names so we can compare them.

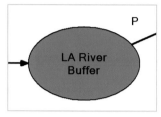

Figure 7-59

Ⓐ In the Catalog window, right-click *Park Suitability Analysis* and choose Edit.

Ⓑ In the Model window, right-click the *LA River Buffer* output data element and choose Model Parameter (Figure 7-59).

We don't have to make a variable first, because data elements are already variables.

Ⓒ Right-click the element again to open its context menu.

The Intermediate status, previously unchecked, is now grayed-out. Once output data is made a model parameter, it *can't* be intermediate.

Ⓓ Click in some empty white space to close the context menu.

Ⓔ Set *Recommended Sites* and *Park Access Zones Pop* as model parameters.

Ⓕ Validate the model, save it, and close it.

Ⓖ In ArcMap, from the main menu, choose Geoprocessing → Geoprocessing Options.

Figure 7-60

Ⓗ In the Geoprocessing Options dialog box, make sure that "Overwrite the outputs of geoprocessing operations" is unchecked.

Ⓘ Make sure that "Add results of geoprocessing operations to the display" is checked (Figure 7-60), then close the dialog box.

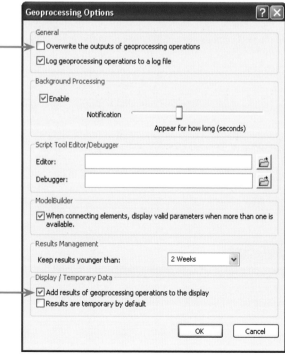

Now that the outputs are tool parameters, they will not be allowed to overwrite feature classes of the same name in the geodatabase. (They won't be allowed to do it when the model is run as a tool. When the model is run from the Model window, it isn't governed by these geoprocessing controls.)

6) Run the tool with different values. Now we'll run the model with a new river buffer value.

Ⓐ In the Catalog window, open the model to run as a tool.

▷ You can also do this by double-clicking its name.

The three new parameters you added show up in the tool dialog box (Figure 7-61). They are marked with error icons 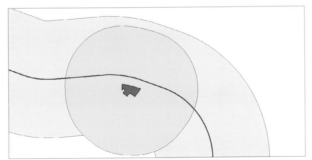 because feature classes with these names already exist in the geodatabase. You can't run the tool until you fix the errors.

Figure 7-61

If ArcMap were set to overwrite inputs in the Geoprocessing Options, you would see a warning icon ⚠ next to these parameters instead. The warning would alert you to change the output names, but you could still run the tool without doing so.

- **B** Change the *LARiverBuffer* output feature class name to **LARiverBuffer2**.

- **C** Change the *RecommendedSites* output feature class name to **RecommendedSites2**.

- **D** Change the *ParkAccessZones_SpatialJoin* output feature class name to **ParkAccessZones2**.

As you click out of each parameter, its warning icon goes away.

- **E** Finally, change the Distance from LA River value to 0.25 miles.

- **F** Compare your dialog box to Figure 7-62, then click OK.

- **G** When the model finishes, close the message box.

You should see three layers in ArcMap. The LA River buffer is skinnier than usual: just a half mile wide now, instead of a mile and a half.

- **H** In ArcMap, drag *RecommendedSites2* to the top of the Table of Contents. If necessary, drag *LARiverBuffer2* to the bottom.

- **I** Optionally, resymbolize the layers and add *LARiver* from the ReadyData geodatabase.

- **J** Zoom in on the suitable site (Figure 7-63).

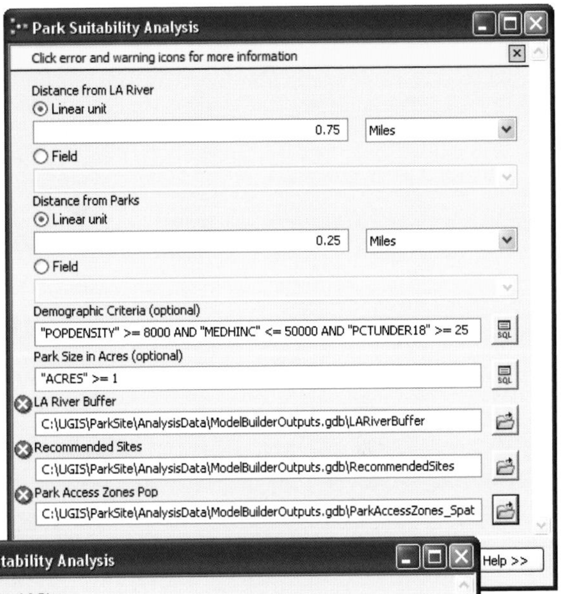

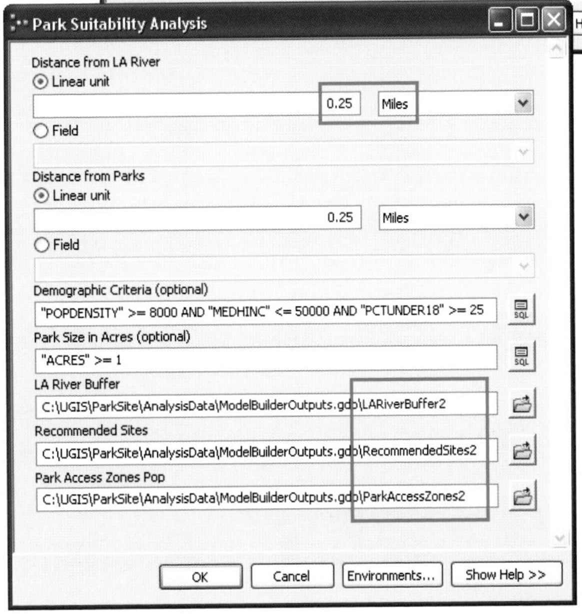

Figure 7-62

Figure 7-63

Using 0.25 miles as the value for distance to the river, we only find one candidate.

K In the Catalog window, expand ModelBuilderOutputs, if necessary.

The geodatabase should contain six feature classes: three from the last run of the model and three from this one.

L Save the map document. Optionally, exit ArcMap.

Exercise 7e: Document the model

If you want people to understand how your model works, you should invest some time in documenting it. The item description is a good place to explain how the model works as a geoprocessing tool. Within the Model window, we can add notes, called labels, to explain what the model is doing at any given point. We can also change the layout to make the model easier to read.

Documenting the model also means keeping a record of when it was run, under what conditions, with what results. One way to do this is with ArcMap's geoprocessing results, which we looked at in Lesson 4. Another way is with a model report.

1) **Edit the item description.** Like datasets, tools have item descriptions. A tool's item description can summarize its purpose and provide usage notes. It can also include keyword tags for finding the tool in searches.

A If necessary, start ArcMap and open Lesson7.

B In the Catalog window, in the Home folder, expand the ParkPlanningModels toolbox. Right-click the model and choose Item Description.

C At the top of the Item Description window, click Edit (Figure 7-64).

Figure 7-64

We won't add an illustration or tags.

D In the Summary box, type: `This model identif ies vacant parcels near the Los Angeles River that may be suitable for a new park.`

▷ Or right-click the model in the Catalog window, choose Properties, click the General tab, and copy and paste the description.

E In the Usage box, type something like this: `Choose a maximum distance from the river, a minimum distance from existing parks, desired neighborhood demographics, and parcel size in acres to find suitable parcels.`

F Scroll down and expand the Distance_from_LA_River parameter.

G For the dialog explanation, type: **The maximum distance a site is allowed to be from the river.** (Figure 7-65)

H Expand the Distance_from_Parks parameter.

I For the dialog explanation, type: **The minimum distance a site must be from an existing park.**

J Optionally, add explanations for the other parameters.

K At the bottom of the Item Description, in the Credits box, type **City of Los Angeles, ESRI.**

L Save the Item description (Figure 7-66), then close it.

Figure 7-65

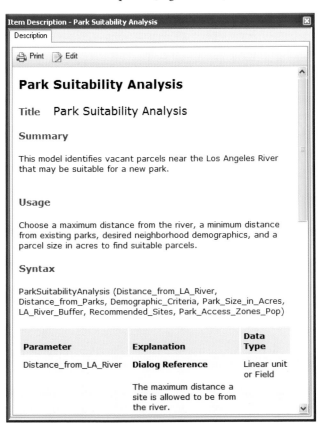

Figure 7-66

2) Change the model layout. The orientation of the model layout can be changed in various ways.

A In the Catalog window, right-click the model and choose Edit.

B Click the Auto Layout and Full Extent buttons.

C From the ModelBuilder main menu, choose Model → Diagram Properties.

Figure 7-67

Figure 7-68

ⓓ On the Diagram Properties dialog box, click the Layout tab.

ⓔ In the Orientation area, click Top to Bottom (Figure 7-67), then click OK.

Nothing happens until you click Auto Layout.

ⓕ Click Auto Layout, then Full Extent (Figure 7-68).

This orientation is better for some purposes, like printing on a vertically oriented page.

ⓖ On the ModelBuilder toolbar, click the Undo button ↶ .

The layout reverts to a horizontal orientation. (The settings don't revert, though: if you click Auto Layout, the model will flow top to bottom again.)

ⓗ Open the Diagram Properties again and click the General tab.

ⓘ In the Layout Mode area, click Manual. Click OK.

The best-looking layout may be one that you design yourself. We'll keep the orientation horizontal, but split it into three tiers. (If you see a way you like better, go ahead and try it. This is just for practice.)

ⓙ On the ModelBuilder toolbar, click the Select tool ➤ .

ⓚ Drag a selection box as shown in Figure 7-69.

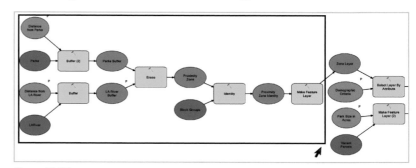

Figure 7-69

When you release the mouse button, the elements covered by the box are selected. You can modify the selection by Control-clicking an element to select or unselect it.

ⓛ Place the mouse pointer over any selected element, then click and drag the entire selection toward the upper left corner of the Model window.

The model remains connected, although a connector arrow may get pretty long.

Ⓜ Drag another selection box as shown in Figure 7-70.

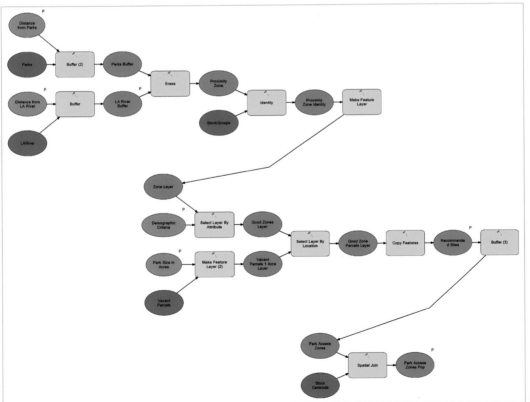

Figure 7-70

Ⓝ Drag the selected elements underneath the first tier of elements and give them a comfortable indent.

Ⓞ Drag a box around the last set of elements and move them underneath the second tier, again with an indent.

Ⓟ Reshape connector arrows as needed by clicking anywhere on them and dragging from the click point to a new location.

Ⓠ Zoom in on the model.

In the best case, the model fits entirely and legibly in your Model window. (You might need to enlarge your window.) The result should look something like Figure 7-71.

Figure 7-71

Ⓡ Click Auto Layout.

You get a message that automatic layout is disabled.

▷ If you didn't get the message, and the layout changed, click the Undo button on the toolbar. Then open the Diagram Properties, click the General tab, click Manual, and click OK.

Ⓢ Click OK on the error message, then save the model.

3) **Label the model.** Labels in a model are helpful in the same way as comment lines in a computer program: they explain what's happening. Labels can be attached to specific elements in the model or they can be free-floating.

Ⓐ From the ModelBuilder main menu, choose View → 100%. Scroll to the upper left corner of the model.

Ⓑ On the Model toolbar, click the Select tool ▶. Hover the mouse pointer over various elements, such as the *Distance from LA River* variable (Figure 7-72).

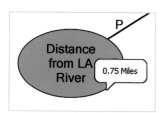

Figure 7-72

A lot of documentation is already built into the tool tips. You see the paths of input and output elements, the values of expressions, and the parameters of tools.

Ⓒ Right-click the *Parks Buffer* output element and choose Create Label.

A label with the default text "Label" is added to the Model window. It may appear on top of a connector arrow.

Ⓓ Click in empty white space to unselect the label and the output data element, then click the label to select it.

Ⓔ Drag the label to a position above the output data element.

As you drag the label, it stays connected to its element by a dotted line. The line disappears when you release the mouse button.

Ⓕ Double-click the label to make it editable.

Ⓖ Type `Sites within park buffers are excluded`.

Ⓗ Place the cursor between the words "park" and "buffers" and press Shift-Enter to break the line. Click in any white space to finish the label (Figure 7-73).

Ⓘ Create labels as suggested below, modifying them as you like:

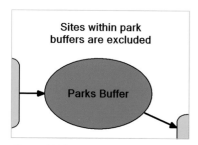

Figure 7-73

- For *LA River Buffer:* `Sites outside this buffer are excluded`.
- For *Proximity Zone:* `Only sites in this area are considered`.
- For the Identity tool: `Adds demographic attributes to proximity zone`.

The labels will look better if we change their font properties.

J Select all four labels by Shift-clicking them.

K Right-click any selected label and choose Display Properties.

The Display Properties window has two columns: on the left are properties that can be set; on the right are the current values of those properties.

L In the Display Properties window, click on the value MS Shell Dlg next to the Font property, then click the button that appears next to the font name (Figure 7-74).

M In the Font dialog box, change the Font Style from Bold to Regular.

N Click the Color drop-down arrow and change the color from Black to Gray.

▷ Or set the font properties to suit yourself.

O Compare your dialog box to Figure 7-75, then click OK.

P Close the Display Properties window and unselect the labels.

The labels you've created so far are connected to specific model elements. You can also create free-floating labels.

Q From the ModelBuilder main menu, choose Insert → Create Label.

R Drag the label to the top center of the Model window. Change its text to something like: **This model finds vacant parcels suitable for park development**.

S Open the label's display properties. Make the font style Regular, the color Gray, and the size 18. Click OK.

T Close the Display Properties window and unselect the label (Figure 7-76).

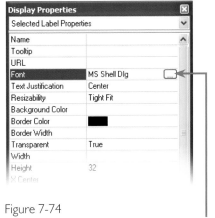

Figure 7-74

Click here

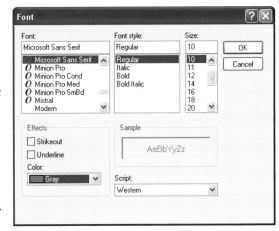

Figure 7-75

Figure 7-76

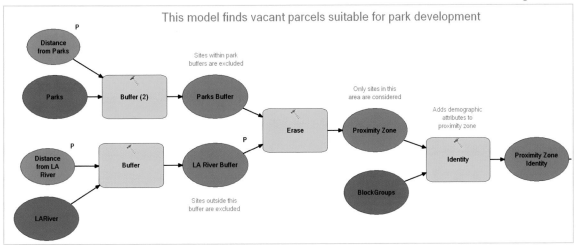

U Optionally, add more labels to the model to explain other processes.

V Zoom to the full extent.

Setting display properties for model elements

To change the appearance of individual model elements (including connector arrows) or groups of selected elements, right-click an element and choose Display Properties. Among the properties you can set are color, shape, size, and text font.

4) View a model report and geoprocessing results. It's important to associate the results of your model with the input conditions that led to them. ModelBuilder can formalize this information in a report.

A From the ModelBuilder main menu, choose Model → Report.

B Click OK to view the report in a window.

The report sorts the model elements into variables and processes (tools).

C In the upper right corner of the Model Report window, click the Expand/Collapse All link (Figure 7-77).

Figure 7-77

D Scroll quickly through the report, then close the window.

By choosing "Save report to a file," instead of "View report in a window," you can save the report to disk as an .xml file.

E Save and close the model.

ArcMap also keeps a log of geoprocessing operations. Since a model is a geoprocessing tool, this gives you another way to get information about the model. By default, results are stored for two weeks, but you can change this on the Geoprocessing Options dialog box.

F From the ArcMap main menu, choose Geoprocessing → Results.

G In the Results window, expand Current Session, then expand the top Park Suitability Analysis entry.

▷ If you started ArcMap at the beginning of this exercise, you'll need to expand Previous Sessions instead of Current Session.

H Expand the Inputs heading (Figure 7-78).

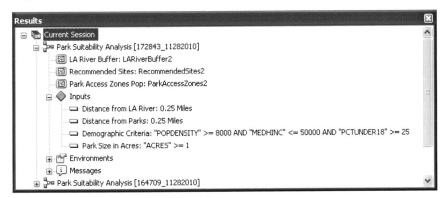

Figure 7-78

The Results window shows you the model's output feature classes, as well as variable values and expressions. You can expand other headings to get information about environment settings and messages.

I Close the Results window.

Model environment settings

In Lesson 4, you set the geoprocessing environment for ArcMap to make sure that output feature classes were projected to the desired coordinate system. You can make environment settings for a model as well. These settings are specific to the model and override environment settings made in ArcMap. If you need to, you can make environment settings that are specific to individual tools within a model. Tool settings override model settings, just as model settings override ArcMap settings.

5) Back up the model. You've tested the model and it works. This is a good time to make a backup copy so you can experiment on your own without worrying that you might break something.

A In the Catalog window, right-click *Park Suitability Analysis* and choose Copy.

B Right-click ParkPlanningModels.tbx and choose Paste.

C Rename the copy in a meaningful way.

▷ You might include the date of the backup in the name (Figure 7-79).

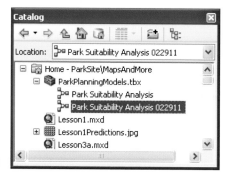

Figure 7-79

The name you give the model in the Catalog window is a label (see page 242). In the model properties, the model's name is ParkSuitabilityAnalysis2. Models are tools, and tools, to satisfy scripting requirements, don't allow spaces in their names.

D Exit ArcMap. Click Yes if you're prompted to save changes.

You can share your model by copying the ParkPlanningTools toolbox to another location. If someone else wants to run the model, they need access to the data and they need ArcGIS Desktop version 10.

You've seen two ways to carry out an analysis project: as a series of stand-alone geoprocessing operations and as a model. Models offer a lot of advantages, but it's an open question as to how fully developed the analysis should be before you turn it into a model.

We recommend planning the analysis in a traditional way: sketching the workflow on paper or a whiteboard, experimenting with geoprocessing tools, and discussing results with colleagues. Wait until you have a pretty firm plan before you take the analysis into ModelBuilder. (And remember that even your "firm plan" will probably change.) If you start with ModelBuilder as your whiteboard, you may find that the model gets too messy in the early planning stages and needs constant corrections. Of course, the more you know about analysis, geoprocessing tools, and ModelBuilder, the sooner you can get your model in good working order. Once that's done, it's relatively easy to add further complexity.

Lesson

8 Present analysis results

COMMUNICATING YOUR FINDINGS

to a diverse audience of readers—with levels of sophistication ranging from public citizens to council members to the City's urban planners, who must ultimately design an actual park—is no easy task.

Our analysis has yielded a short list of candidate sites that we want to present to the city council and other interested parties. We'll make our presentation in the form of a map that places the results in a meaningful geographic context, that addresses the project guidelines, and that follows good cartographic design principles.

The deliverable item is an 8½ x 11-inch map that can be printed or viewed on-screen.

But wait, you may be thinking, what's wrong with simply adding the layer of recommended sites to an online basemap, like the imagery or street basemaps we've used throughout the book? There's nothing wrong with it, and basemaps will be part of our map design, but our map design needs to accomplish a lot of things. We'll be adding inset maps, customizing the positions of labels and other text, adding graphs of statistical data, and incorporating other standard cartographic elements like scale bars and north arrows.

Our analysis results aren't too complex. We basically just want to show the study area, represented by the LA River buffer zone, and the five proposed park boundaries. We want to set these analytical layers against a backdrop of topography, place-names, and major roads. Perhaps the biggest design challenge we face is that these five locations are small parcels of land strung out over a 20-mile stretch of river corridor: how do we map the area of interest without rendering the parks too tiny to see at that scale? The solution lies in the use of inset maps—smaller maps placed within the main map, each one portraying a close-up view of a particular site.

Lesson Eight roadmap

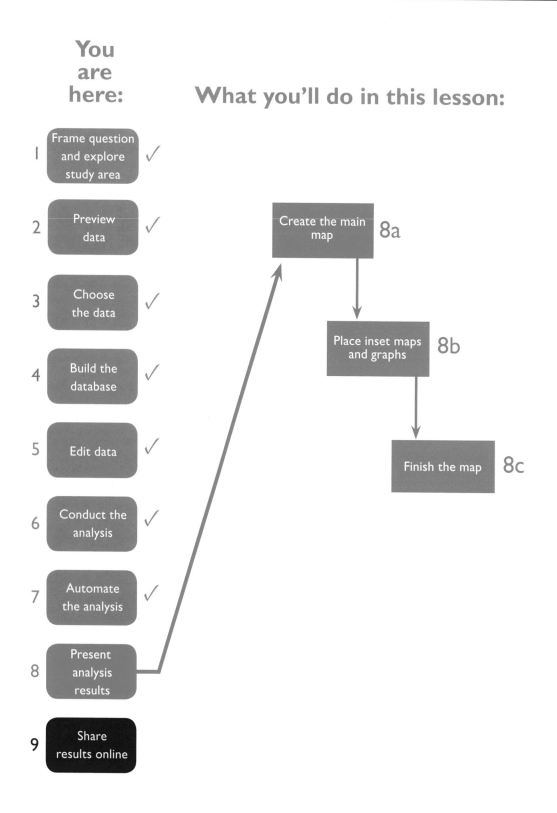

You are here:

1 Frame question and explore study area ✓

2 Preview data ✓

3 Choose the data ✓

4 Build the database ✓

5 Edit data ✓

6 Conduct the analysis ✓

7 Automate the analysis ✓

8 Present analysis results

9 Share results online

What you'll do in this lesson:

Create the main map — 8a

Place inset maps and graphs — 8b

Finish the map — 8c

Another challenge we'll face is how to incorporate comparisons of some of the key demographic variables that drove the selection process: population density, median household income, and the percentage of children.

Cartography involves innumerable choices about color, font, line width, the placement and juxtaposition of elements, and so on. These are often somewhat subjective decisions, with many good (and bad!) possibilities. This lesson takes you through the creation of an exact design (seen in Figure 8-1 at 60% actual size), but along the way you'll have the option of altering things to suit your own taste. We'll try to explain why we made the choices we did, but you don't have to follow the instructions with total fidelity. Experimenting with some of the settings will give you a deeper understanding of how the software works, and a chance to make a map that suits your own aesthetic tastes.

Figure 8-1

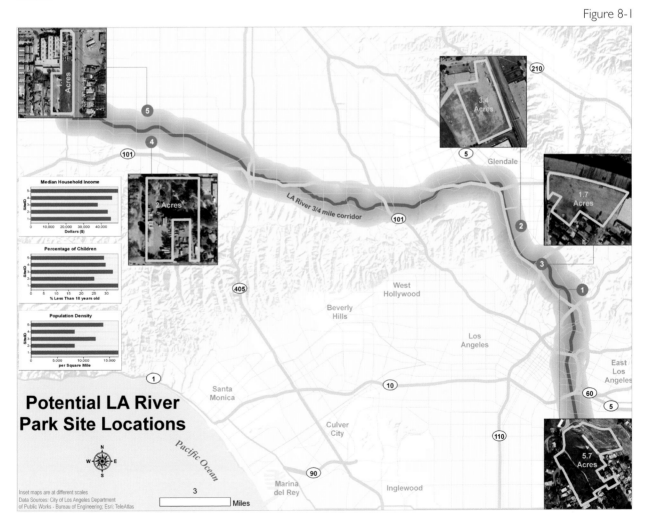

A full presentation of this type in the real world would be likely to include many different media elements: large format wall maps, smaller notebook-sized maps, printed reports, digital slide shows, PDF downloads, and websites. We can't do all these things without writing another book, so we're going to focus on a single map layout that can be printed on a standard letter-sized piece of paper. In Lesson 9, we'll go a step further and make a web map, but for now, our working medium will be a sheet of letter-sized paper.

In reality, an 8½ by 11-inch map is pretty small, but it would be worse if we were trying to pack a lot more features or a lot of detailed data into this format. We're using this small size in recognition of the fact that many people don't have access to larger format printing devices. And if you can design well in small format, you'll feel that much more comfortable and creative when you begin working in larger formats.

A good way to begin a map layout—just like an analysis—is with a sketch of your basic plan. The sketch for our map is shown in Figure 8-2:

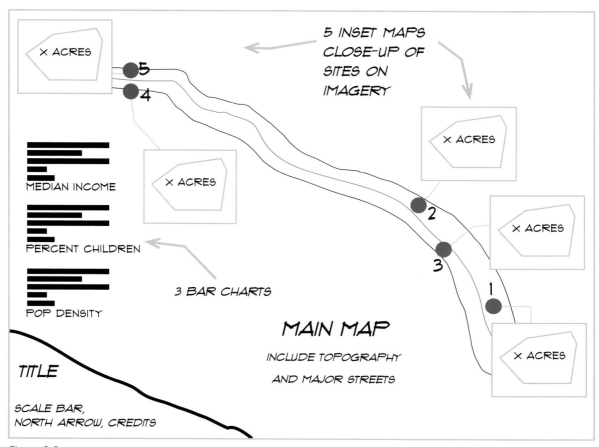

Figure 8-2

Exercise 8a: Create the main map

Beginning with this exercise, we'll be working in what is called *layout view* in ArcMap. Viewing and working with data is one thing; composing a map for the printed page is something else. In composing maps you must take into account not only your specific design considerations but also output constraints such as paper size, page orientation, and print margins. Map composition is done in layout view, the graphical, page-layout environment of ArcMap.

1) Open a map template. We'll start with a new map document created from a standard template—in this case a landscape (horizontally oriented) page.

Ⓐ Start ArcMap.

Ⓑ In the Getting Started dialog box, click Templates in the left-hand window. Under Standard Page Sizes, click North American (ANSI) Page Sizes.

Ⓒ In the right-hand window, scroll down and click Letter (ANSI A) Landscape to select it (Figure 8-3), then click OK.

Even if your printer uses different sizes, use ANSI A anyway to complete the exercise.

Choosing the template makes ArcMap open in layout view (Figure 8-4). You can tell because you're looking at an empty data frame on a representation of a piece of paper. Also, a new toolbar, the Layout toolbar, should appear on the interface.

Select North American Letter Size 8 1/2" x 11"

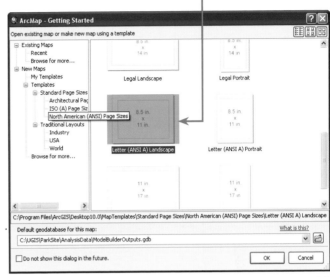

Figure 8-3

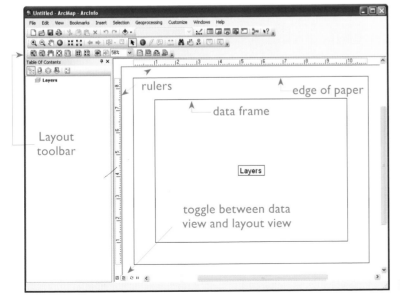

Figure 8-4

D If you don't see the Layout toolbar, from the ArcMap main menu choose Customize → Toolbars and check Layout.

The idea of a data frame takes on new significance in layout view. In data view (where we've worked until now), a data frame is a more or less abstract "container" of layers. Apart from one or two special properties, like its coordinate system setting that controls on-the-fly projection, we generally don't pay too much attention to it. In layout view, the data frame is visual: it's a rectangular window placed on a virtual sheet of paper and a key element in the map composition.

In the template, the default margins around the data frame are set to one inch on all sides. Most of this space would be wasted in our layout because we're going to place all the map content—data and marginalia both—directly in the data frame. We'll enlarge the data frame, leaving just a ¼-inch margin around the edge of the page.

E From the ArcMap main menu, choose Customize → ArcMap Options.

F In the ArcMap Options dialog box, click the Layout View tab.

G In the "Grid" area, check the Show box and confirm that the horizontal and vertical spacing values are set to 0.25 inches.

▷ If necessary, choose ¼ in from the drop-down lists.

H In the "Snap elements to" area, check the Grid box.

I Confirm that your settings match Figure 8-5, then click OK.

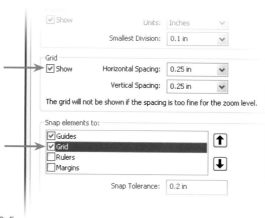

Figure 8-5

A pattern of gray dots appears on the virtual page. This is the grid. The outermost dots mark the quarter-inch margin on all sides.

J On the Tools toolbar, click the Select Elements tool ▸ if necessary. Click anywhere inside the data frame to select it.

K Resize the data frame by dragging its blue selection handles.

As you drag the data frame, it snaps to the dots on the grid. Your result should look like Figure 8-6. We don't need the grid anymore, so we'll turn it off.

L From the main menu, choose Customize → ArcMap Options.

M In the "Grid" area, uncheck the Show box.

N In the "Snap elements to" area, uncheck the Grid box. Click OK.

O Click anywhere outside the data frame to unselect it.

2) **Add the LA River to the layout.** The *LARiver* feature class, like all our geodatabase data, is in the State Plane California Zone 5 coordinate system. This is the system we want our map to use. By adding *LARiver* as the first layer, we'll set the data frame to this system and ensure that any basemap layers added thereafter will be projected on the fly to match.

A Maximize the ArcMap window.

B On the Layout toolbar, click the Zoom Whole Page button 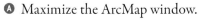.

C In the Catalog window, navigate to C:/UGIS/ParkSite/ AnalysisData/ReadyData.gdb.

D Drag and drop *LARiver* onto the map.

  ▷ We want to maximize our layout space, so keep the Catalog window in its Auto Hide (unpinned) state.

E In the Table of Contents, open the Symbol Selector for the *LARiver* layer.

F In the Symbol Selector, click the predefined River symbol.

G Change the width to 4 points, then click OK (Figure 8-7).

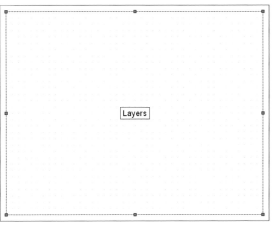

Figure 8-6

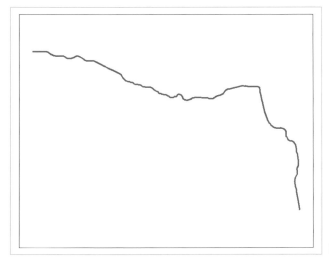

Figure 8-7

This is a good time to make the distinction between the navigation tools on the Layout toolbar (Figure 8-8) and their counterparts on the Tools toolbar.

The icons for navigation tools on the Layout toolbar show a "page" in the background.

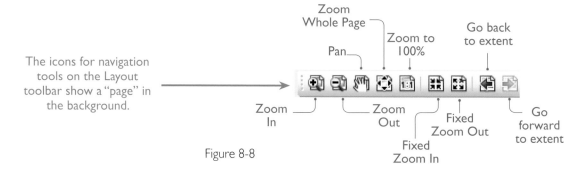

Figure 8-8

The layout navigation tools magnify, reduce, and otherwise adjust your view of the virtual page. They do not change the scale or extent of the map within the data frame. Both sets of tools work in layout view, so you need to be careful. (Later on, when we've finalized the map scale and extent, we'll disable the data navigation tools.)

🅗 Note the map scale in the scale box on the Standard toolbar and note the view scale on the Layout toolbar.

🅘 On the Layout toolbar, click the Zoom In tool 🔍 and zoom in to the upper left corner of the data frame. It doesn't matter exactly how far.

Notice the effect in Figure 8-9. The rulers have zoomed in, showing you that you're now seeing just a few inches of the page. The line feature also got noticeably thicker. The map scale hasn't changed within the data frame, but the view scale is different.

🅙 On the Layout toolbar, click the Go back to extent button 🔙.

🅚 Now, on the Tools toolbar, click the Zoom In tool 🔍 and zoom in on the same upper left corner of the data frame.

▷ You have to begin drawing your zoom rectangle *within* the data frame or the tool won't have any effect.

This time the map scale changes (you've zoomed in on the data) and the view scale stays the same.

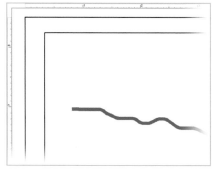

Figure 8-9

🅛 In the Table of Contents, right-click the *LARiver* layer and choose Zoom To Layer.

3) Add a terrain basemap. The Terrain basemap will make an unobtrusive background, with topographic relief that will make the map less starkly flat.

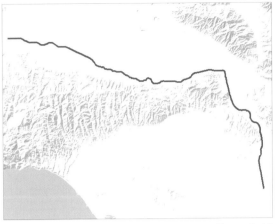

Ⓐ Add the Terrain basemap (Figure 8-10).

▷ If you get a Geographic Coordinate Systems Warning, check the "Don't warn me again in this session" box and close the warning.

Before going on, we'll give the map a title—which is different from its file name—and save the map document.

Ⓑ From the main menu choose File → Map Document Properties.

Ⓒ In the Title box, type **Potential LA River Park Site Locations** (Figure 8-11).

Figure 8-10

Figure 8-11

Ⓓ Optionally, fill out the other map document properties as you did in Lesson 6 (page 229).

Ⓔ Check the box to store relative pathnames to data sources.

Ⓕ Click Make Thumbnail, then click OK.

Ⓖ On the Standard toolbar, click the Save button 💾.

Ⓗ In the Save As dialog box, navigate to \UGIS\ParkSite\ AnalysisData and save the map document as **Lesson8**.

We're saving the map document in the AnalysisData folder, instead of MapsAndMore, because it belongs to our final project results, and we want to store it with the starting and output data.

4) Add and symbolize the LA River buffer. In Lesson 6, we created the *LARiverBuffer* feature class. We'll add it now to show the study area, and symbolize it with a special gradient fill symbol.

Ⓐ In the Catalog window, under AnalysisOutputs.gdb, add *LARiverBuffer* to the map.

Ⓑ In the Table of Contents, open the Symbol Selector for the *LARiverBuffer* layer.

G Click Edit Symbol to open the Symbol Property Editor.

We opened the Symbol Property Editor once before in Lesson 1 to add a halo to a label. That earlier effect and the gradient fill we're about to use now barely scratch the surface of what you can do with symbols. See the topic *Editing symbol properties* on page 295 for more information.

Right-click to toggle off Graphic View.

Figure 8-12

Select this color ramp (Green Bright).

D In the Symbol Property Editor, click the Type drop-down arrow and choose Gradient Fill Symbol.

E On the Gradient Fill tab, click the Style drop-down arrow and choose Buffered.

F Set the Intervals to **40**.

G Set the Percentage to **20**.

H Click the Color Ramp drop-down arrow and choose Green Bright (Figure 8-12).

▷ Or right-click the color ramp, uncheck Graphic View, and choose the ramp by name.

I Click OK to close the Symbol Property Editor.

J In the Symbol Selector, set the outline color to No Color.

K Click OK to close the the Symbol Selector.

A gradient fill changes color or intensity across the symbol. Of the four styles, we chose Buffered because it makes the pattern radiate outward along the length of the river. The number of intervals determines the smoothness of the effect: with more intervals, color transitions are smoother but the symbol takes longer to draw. The percentage controls the pattern of color saturation. You may want to experiment with different settings or ramps as there are no strict rules at work. The preview in the upper left corner of the Symbol Property Editor helps you anticipate the effects of your settings.

Figure 8-13

L Open the layer properties for the *LARiverBuffer* layer and click the Display tab.

M In the Transparent % box, type **30**, then click OK.

The result is a smooth gradient fill with terrain showing through around the outer edges of the buffer (Figure 8-13). It looks good, but it also has a subtle analytic effect. Our analysis prefers candidates closer to the river, and the symbol is a visual reminder that a candidate in a dark green part of the zone is closer to the river than a candidate in a light green part.

Why make the buffer green? One reason is that we have many layers to symbolize and green works well in combination with the other colors we're going to use. On a psychological level, however, we're also suggesting that this area is verdant, riparian habitat—whereas, in reality, much of it is either densely developed residential/commercial land or industrial blight. Are we misleading our audience or just being persuasively optimistic? Whichever way you think of it, you need to be aware as a cartographer that symbols may have hidden meanings as well as overt ones.

5) Set basemap transparency. In preparation for the next step, we'll make the terrain basemap partially transparent.

Ⓐ Open the layer properties for the *Basemap* layer and click the Display tab.

Ⓑ In the Transparency % box, type **40** (Figure 8-14), then click OK.

The transparency value is applied to the group layer as a whole, although only the *Terrain* sublayer is turned on. You can also set certain properties by opening layer properties at the sublayer level.

On the map, the only difference you might notice is that the *Basemap* layer is a little bit lighter. The transparency effect won't become obvious until you add some data below it.

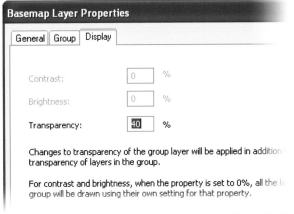

Figure 8-14

6) Add and symbolize the LA city limits. We only considered locations within Los Angeles itself, so it will be helpful to show the city limits on the map.

Ⓐ In the Table of Contents, collapse the *Basemap* layer.

Ⓑ In the Catalog window, under ReadyData.gdb, drag *Cities* to the Table of Contents and drop it at the bottom, underneath the *Basemap* layer (a horizontal black bar marks the spot).

When you drop a layer on the map, it takes its "natural" place in the top-to-bottom order (point, line, polygon, raster). When you drop it in the Table of Contents, it goes exactly where you put it. It may seem counterintuitive to place a layer of features below a basemap, but doing so will allow us to create a nice, unobtrusive effect.

Ⓒ Open the layer properties for the *Cities* layer and click the Definition Query tab.

Ⓓ Click Query Builder.

Ⓔ In the Query Builder dialog box, create the following expression:

```
"NAME" = 'Los Angeles'
```

SELECT * FROM Cities WHERE:
"NAME" = 'Los Angeles'

Figure 8-15

This is the very same definition query we created at the beginning of Lesson 1.

ⓕ Check your query against Figure 8-15, then click OK on the Query Builder dialog box.

ⓖ On the Layer Properties dialog box, click Apply, then click the Symbology tab.

ⓗ On the Symbology tab, click the symbol patch to open the Symbol Selector.

ⓘ Set the fill color to Yucca Yellow (Figure 8-16) and the outline color to No Color.

ⓙ Click OK on both the Symbol Selector and the Layer Properties dialog box.

The result is a subdued yellow filtered through the semi-transparent basemap. White space shows through areas that aren't part of Los Angeles proper. Some of these areas are pretty famous places, like Beverly Hills and Santa Monica, and we'll label them in Exercise 8c.

7) **Add and symbolize the LA County boundary.** The white showing through is a bit stark, so we'll mute it by putting the *Counties* layer underneath *Cities*.

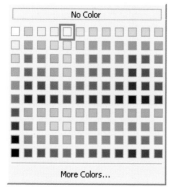

Figure 8-16

ⓐ In the Catalog window, under ReadyData.gdb, drag *Counties* to the bottom of the Table of Contents, underneath the *Cities* layer.

ⓑ Open the Symbol Selector for the *Counties* layer.

ⓒ Set the outline color to No Color.

ⓓ For the fill color, at the bottom of the color palette, click More Colors to open the Color Selector.

In the Color Selector, you can mix your own colors using either the RGB or CMYK color models. We'll work with RGB, which is the model used for computer monitors. (The HSV model, which is similar to RGB, is also available.) When you design a map for printed output, you have to expect that there will be color shifts. Since these shifts vary according to the output device or printing process, you should make test prints of your map as it develops and modify the on-screen colors as needed.

Figure 8-17

Type in values

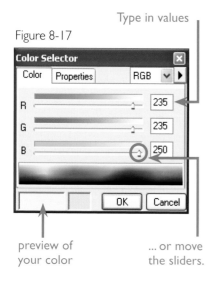

preview of your color

... or move the sliders.

ⓔ In the Color Selector, make sure the Color tab is selected and the drop-down list is set to RGB.

ⓕ Type the following values: R = **235**, G = **235**, B = **250** (Figure 8-17). Click OK on the Color Selector and the Symbol Selector.

On the map, you should see a subtle contrast between the City of Los Angeles, the other land area, and the water of the Pacific Ocean (Figure 8-18).

Editing symbol properties

When you add a layer to ArcMap, it's symbolized with a "simple" symbol type. Point layers are symbolized with simple marker symbols, polylines with simple line symbols, and polygons with simple fill symbols. Simple symbol types have just a few editable properties, like size, width, and color, which can be changed on the Symbol Selector. To make more complex symbols, open the Symbol Property Editor by clicking Edit Symbol on the Symbol Selector, and change the symbol type at the top of the dialog box. (The predefined symbols that you choose in the Symbol Selector are often non-simple types.)

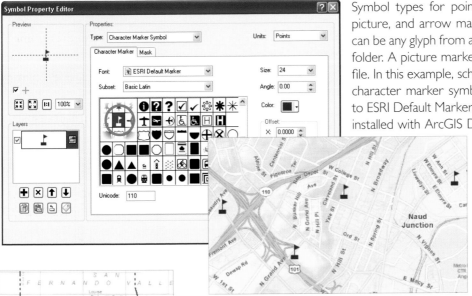

Symbol types for point layers include character, picture, and arrow markers. A character marker can be any glyph from any font in your system font folder. A picture marker can be any .bmp or .emf file. In this example, schools are symbolized with a character marker symbol. The font has been set to ESRI Default Marker, one of many symbol fonts installed with ArcGIS Desktop. The symbol is the glyph circled in blue.

To look for predefined symbols, like this one for schools, type a keyword in the search box at the top of the Symbol Selector.

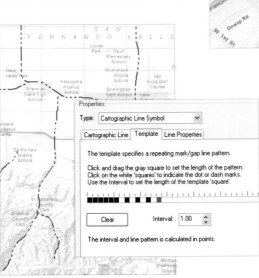

Symbol types for polygon layers include line, marker, picture, and gradient fills. A line fill symbol fills polygons with lines of any width, spacing, angle, and color. You can also fill polygons with markers or pictures. A gradient fill symbol fills polygons with a color ramp. Here, a line fill symbol is used for a buffer zone along the LA River. The symbol is built from two superimposed graphic "layers."

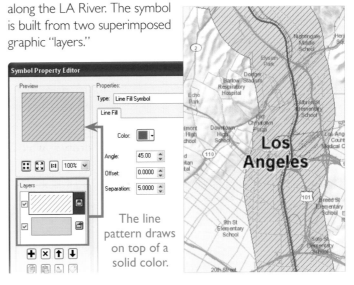

The line pattern draws on top of a solid color.

Symbol types for polyline layers include cartographic, hash, marker, and picture. Cartographic line symbols let you make dash patterns and other effects. Marker and picture line symbols let you place recognizable icons at intervals along a line (to symbolize a bus or bike route, for example). Here, a cartographic line symbol is used to make a dash pattern for intermittent streams.

8) Add and symbolize major roads. Road networks strongly influence our sense of local geography and are important for orientation on many maps.

Figure 8-18

A In the Catalog window, under ReadyData.gdb, add *MajorRoads* to the map.

  ▷ It should appear at the top of the Table of Contents. If it doesn't, move it there now.

B Open the attribute table for the *MajorRoads* layer and scroll across the attributes.

The table contains the name of each road and various other descriptors.

C Right-click the ROAD_TYPE field heading and choose Sort Ascending (Figure 8-19).

D Scroll down through the table and note the values in this field.

There are three road type values. "Highway" is for secondary state and county highways; "Limited Access" is for big freeways and interstate

Table

MajorRoads

	PREFIX	PRETYPE	NAME	TYPE	SUFFIX	CLASS	HWY_TYPE	HWY_SYMBOL	ROAD_TYPE	Shape_Length
	E		Telegraph	Rd		2	S	126	Highway	35544.351558
		Sthy	126			2	S	126	Highway	7283.145682
			Pacific Coast	Hwy		2	S	1	Highway	27539.314676
			Pacific Coast	Hwy		2	S	1	Highway	17939.909179
			Pacific Coast	Hwy		2	S	1	Highway	151.229247
			Pacific Coast	Hwy		2			Highway	488.085619
			Pacific Coast	Hwy		2			Highway	3071.332619
		Sthy	138			2	S	138	Highway	431.782257

|◄ ◄   0 ► ►| (0 out of 2355 Selected)

MajorRoads

Figure 8-19

highways; and "Road" typically designates important streets. We'll symbolize the layer on the values in this field, using different line thicknesses.

Take note also of the HWY_SYMBOL attribute next to ROAD_TYPE. This field stores numeric designators, primarily for the freeways and highways. We'll use this field to label the roads in Exercise 8c.

E Close the attribute table.

F Open the layer properties for the *MajorRoads* layer. Click the Symbology tab, if necessary.

G In the Show box, click Categories and accept the default Unique Values option.

H Set the Value field to ROAD_TYPE.

I Click Add All Values (Figure 8-20).

J In the Symbol column, double-click the Highway symbol to open the Symbol Selector.

K Set its color to Gray 20% and its width to 2. Click OK.

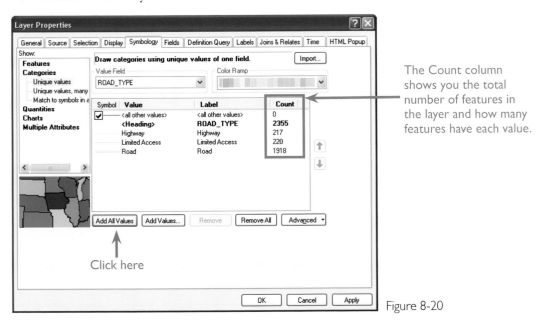

The Count column shows you the total number of features in the layer and how many features have each value.

Figure 8-20

L For the Limited Access symbol, set the color to Gray 20% and the width to 4.

M For the Road symbol, set the color to Gray 20% and the width to **0.5**.

N In the Value column, click Limited Access to highlight it, then click the Up arrow on the right side of the dialog box.

In the Table of Contents, the values will now be listed in descending order of thickness, which looks good. Manipulating this order, however, has no effect on which features draw on top of which other features in the same layer.

○ In the Symbol column, uncheck the box for <all other values>.

Since there aren't any other values (the Count is 0), the map will look the same whether or not the box is checked. The point of unchecking the box is to prevent the <all other values> label from showing up in the Table of Contents.

Symbol levels

Sometimes you want to control the drawing order of features within a layer. For example, suppose you have a *Roads* layer that includes paved roads, unpaved roads, and foot trails. If these values are symbolized with distinct colors or dash patterns, it will be obvious, wherever features of different types cross, that one is on top and one is underneath. You might want paved roads to draw on top of unpaved roads and unpaved roads to draw on top of trails. You can control this feature drawing hierarchy within a layer by setting symbol levels. On the Symbology tab, click the Advanced button to access the Symbol Levels dialog box. (We avoided this problem with the *MajorRoads* layer because all three road types have the same symbol color.)

● Check your settings against Figure 8-21, then click OK.

The *MajorRoads* layer adds context to the map without overwhelming it (Figure 8-22) and the symbology is fairly self-explanatory.

Figure 8-21

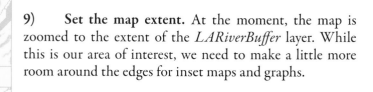

Symbol	Value	Label	Count
☐	<all other values>	<all other values>	0
	<Heading>	**ROAD_TYPE**	**2355**
	Limited Access	Limited Access	220
	Highway	Highway	217
	Road	Road	1918

Use the arrows to move a highlighted value up or down.

Figure 8-22

9)   **Set the map extent.** At the moment, the map is zoomed to the extent of the *LARiverBuffer* layer. While this is our area of interest, we need to make a little more room around the edges for inset maps and graphs.

Ⓐ On the Standard toolbar, set the map scale to **1:160,000.**

At this scale, one inch on the map equals about two and a half miles on the ground. With your ArcMap window maximized, the map extent should be close to Figure 8-23, although the result depends on things like the size and screen resolution of your monitor. For our purposes, the extent is more important than the scale.

Ⓑ If necessary, use navigation tools on the Tools toolbar (not the Layout toolbar!) to approximate the map extent in the figure.

Now we'll lock this extent so we don't change it by mistake.

Ⓒ Open the data frame properties and click the Data Frame tab.
  ▷ How? In the Table of Contents, right-click the name of the data frame (Layers) and choose Properties.
Ⓓ On the Data Frame tab, click the Extent drop-down arrow and choose Fixed Extent (Figure 8-24).
Ⓔ Click the Specify Extent button (Figure 8-25).

The default setting, Current Visible Extent, is what we want.

Ⓕ Close the Data Frame - Fixed Extent dialog box.
Ⓖ Click Apply on the Data Frame Properties dialog box, but leave it open.

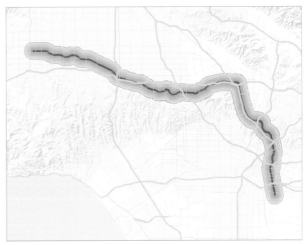

Figure 8-23

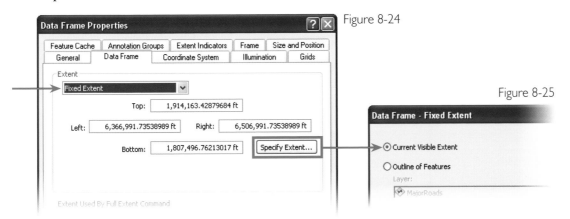

Figure 8-24

Figure 8-25

  ▷ If you get a coordinate system warning, check the "Don't warn me again in this session" box, then click Yes.

H Click the General tab and change the name of the data frame from Layers to **Main Map**. Click OK on the Data Frame Properties dialog box.

In the Table of Contents, the data frame is renamed. On the Tools toolbar, the navigation tools are disabled because you fixed the extent. You can't zoom or pan the map.

10) Add and symbolize recommended sites. The five recommended sites are the most important features on our map, but they're so small and spread out that we need to do some creative symbology to make them show up. In our sketch at the beginning of the lesson, we used red circles to represent the sites. We'll create that effect here by symbolizing the *RecommendedSites* layer (a polygon layer) with graduated symbols. We used this technique before in Lesson 1 (pages 46–48) to show income values within block groups as circles of varying size. Our purpose here is more unorthodox. Because we can easily enlarge the symbols to a size that will draw the map reader's attention, we'll use them as substitutes for the small site polygons. It will still be important to show the actual shape and size of the sites, and to do that we'll create inset maps in the next exercise.

Figure 8-26

Add this
layer file

A In the Catalog window, expand Folder Connections, if necessary, and navigate to the MapsAndMore folder.

B Add the layer file *RecommendedSites.lyr* to the map (Figure 8-26).

▷ If you don't have this file, you can download it from the Lesson 6 results at the Understanding GIS Resource Center.

In the Table of Contents, the layer appears directly above the *LARiverBuffer* layer. It has the bright green outline and hollow fill we gave it in Lesson 6.

C Open the attribute table for the *RecommendedSites* layer.

D Confirm that it shows the field aliases and formatting from Lesson 6, then close the table.

The point of using the layer file is not for its symbology (which we're about to change) but for its attribute formatting. We'll take advantage of this in the next exercise when we work with graphs and inset maps.

E Open the layer properties for *RecommendedSites* and click the Symbology tab, if necessary.

F In the Show box, click Quantities. Under Quantities, click Graduated Symbols.

G In the Fields area, click the Value drop-down list and choose SITEID.

Each of the five sites is assigned a symbol. By default, the symbols increase in size from 4 to 18 points. That would be fine if we were representing a range of attribute values, but it doesn't suit our special case. We want all our "graduated" symbols to be the same size.

H Click the Template button to open the Symbol Selector.

I In the list of symbols, click Circle 1.

J Make the color Mars Red (Figure 8-27), then click OK on the Symbol Selector.

K On the Symbology tab, in the Symbol Size boxes, replace both the "from" and "to" values with **22**.

L Check your settings against Figure 8-28, then click OK.

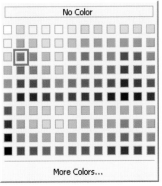

Figure 8-27

Figure 8-28

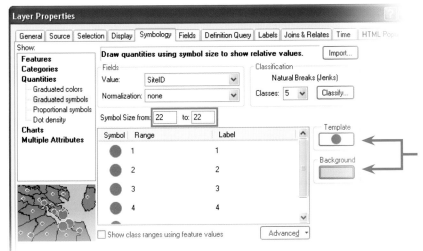

The red circle would normally draw inside the background (the polygon fill color), but in this case, the symbol will completely cover the features at our map scale.

On the map, the symbols stand out prominently (Figure 8-29). Cartographers often use very bright colors (called spot colors) for the features they most want to call attention to on the map.

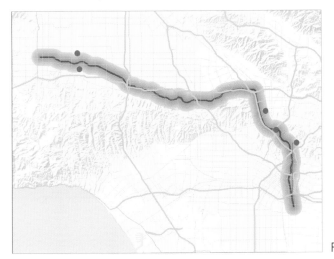

Figure 8-29

11) Label the sites. We'll use the SITEID field again to label the sites.

- Ⓐ Open the layer properties for *RecommendedSites* and click the Labels tab.

- Ⓑ In the upper left corner of the dialog box, check the "Label features in this layer" box.

- Ⓒ Click the Label Field drop-down arrow and choose SITEID.

- Ⓓ In the Text Symbol area, accept the default font (Arial), but change the size to 9 and click the Bold button **B**.

- Ⓔ Set the color to Arctic White.

- Ⓕ Check your settings against Figure 8-30, then click OK.

Figure 8-30

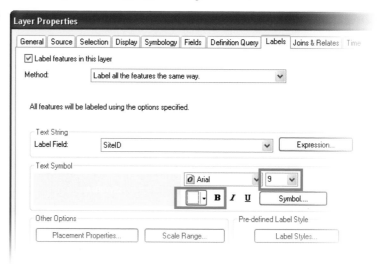

On the map, the symbols are labeled with their SITEID numbers.

- Ⓖ On the Layout toolbar, click the Zoom In tool 🔍 and zoom in on the three sites shown in Figure 8-31 for a closer look.

Figure 8-31

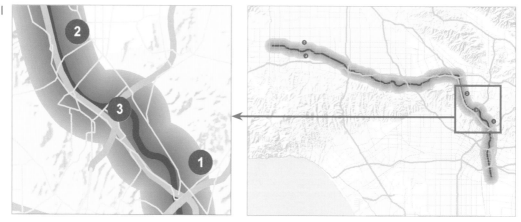

H Click the Zoom Whole Page button ⊡ to return to the full page view.

I In the Table of Contents, collapse the *MajorRoads* and *RecommendedSites* layers so they don't take up so much room.

J Save the map.

K If you're continuing to the next exercise now, leave ArcMap open; otherwise, exit ArcMap.

In this lesson, we've symbolized the five layers that make up the main map. In the next exercise, we'll add inset maps to show close-ups of the sites and graphs that compare the neighborhood demographics.

Exercise 8b: Place inset maps and graphs

An inset map is a small map positioned within a large map. It typically shows a part of the main map at a larger scale to reveal more detail, which is what our inset maps will do. (Inset maps can also be used in the opposite way: to give a small-scale overview of the region within which the main map is situated.)

We want map readers to have a close-up look at each of the sites, so we'll create five large-scale inset maps. We can't simultaneously set different map scales within a single data frame, (except with viewer windows), so we need to insert some new data frames. In fact, we'll insert five of them and place them around the main map as shown in our layout sketch. After we finish the inset maps, we'll add bar graphs comparing median income, the percentage of children, and population density for the sites.

1) Insert a data frame. We'll insert a new data frame to contain our first inset map.

A If necessary, start ArcMap and open Lesson8 from the list of recent maps.

B From the main menu, choose Insert → Data Frame.

Figure 8-32

A new data frame is placed in the center of the map display (Figure 8-32) and selected. In the Table of Contents, its name is boldfaced to show that it's the active data frame.

C Open the data frame properties for New Data Frame.

D In the Data Frame Properties dialog box, click the Size and Position tab.

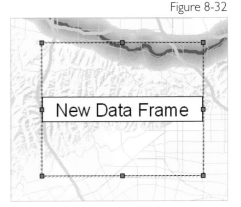

Figure 8-33

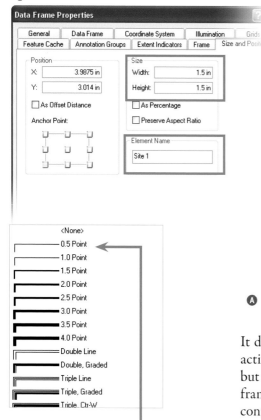

Choose the 0.5 point
border.

E In the Size area, set the Width to **1.5** and the Height to
1.5.

The values default to inches (in).

F In the Element Name box, rename the data frame to
Site 1.

Don't worry about the X and Y position values: we'll
position the data frame manually.

G Click the Frame tab.

H Click the Border drop-down arrow and click the 0.5
Point border.

I Check your settings against Figure 8-33 and click OK.

On the map, the data frame is resized and renamed.

2) Add the recommended sites. We'll add the
RecommendedSites layer to each new data frame, then zoom
in on the various features of interest.

A In the Catalog window, under MapsAndMore, drag and drop
RecommendedSites.lyr onto the layout.

It doesn't matter where you drop it—the layer will be added to the
active Site 1 data frame. The layer is listed in the Table of Contents,
but on the map the features are barely visible in the transparent data
frame. (The data frame label goes away now that the data frame has
content.)

B Open the attribute table for the *RecommendedSites* layer in the
Site 1 data frame.

▷ Make sure you're not opening it from the Main Map data frame.

C If necessary, move the table away from the Site 1 data frame.

D In the attribute table, right-click the small gray box next to the
first record (SITEID 1) and choose Zoom To.

On the map, the feature fills the data frame (Figure 8-34).

E Close the attribute table.

Right-click
here

Figure 8-34

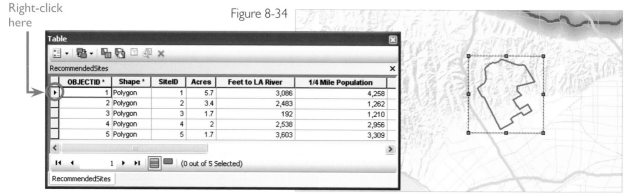

3) Add basemap imagery to the inset map. At the large scale of the inset map, imagery will show the site as it really looks. (Keep in mind that basemap layers are updated periodically, so your imagery may look different.)

Ⓐ Add the *Imagery* basemap to the Site 1 data frame.

The feature symbology is good, but we'll make it just a little more attention-getting.

Ⓑ In the Table of Contents, under the Site 1 data frame, open the Symbol Selector for *RecommendedSites*.

Ⓒ Change the Outline Width to 3 and the Outline Color to Solar Yellow (Figure 8-35), then click OK.

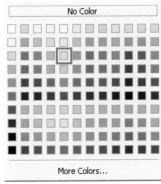

Figure 8-35

Your inset map should look like Figure 8-36.

4) Label the site. We want to label the recommended sites with their sizes so the map reader knows how big they are. At the same time, as we get closer to finishing the map, we're also going to be more particular about the appearance and placement of text. ArcMap's standard dynamic labeling has served us well, but at this point we're ready to give up some speed and simplicity in order to gain more control.

ArcMap has an advanced labeling "engine" called Maplex, which we'll start using now. Like the standard label engine, Maplex places labels according to rules, and updates their positions dynamically, but because its rules are more sophisticated, it gives us more of a hand in the process and generally produces better results. Eventually, we'll want to take complete manual control of text placement and we'll see how to do that in Exercise 8c.

Figure 8-36

Ⓐ On the main menu, choose Customize → Extensions.

Ⓑ In the Extensions dialog box, check the the Maplex box (Figure 8-37), then click Close.

Ⓒ Open the data frame properties for Site 1 and click the General tab.

Ⓓ Near the bottom of the dialog box, click the Label Engine drop-down arrow and choose ESRI Maplex Label Engine (Figure 8-38). Click OK on the Data Frame Properties dialog box.

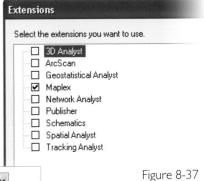

Figure 8-37

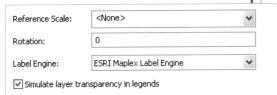

Figure 8-38

Maplex is one of several extension products that enrich the core functionality of ArcGIS Desktop. It improves the placement of labels with advanced rules that include the stacking and curving of text and adjustment of font size. Starting with release 9.1 of ArcGIS Desktop, Maplex is included at no cost with an ArcInfo license. (ArcView or ArcEditor users have to license Maplex separately.) Because Maplex is more computationally intensive than the standard label engine, you may want to use it in some data frames and not others. To make Maplex available, you turn it on in the Extensions dialog box. The functionality is not actually used until you turn it on in the data frame properties.

If a user without a Maplex license opens a map document in which the Maplex engine is used, the labels will lose their Maplex-specific properties and revert to the placement properties of the standard label engine.

ⓔ Open the the layer properties for *RecommendedSites* and click the Labels tab if necessary.

ⓕ Check the "Label features in this layer" box. For the moment, ignore the Label Field setting.

ⓖ In the Text Symbol area, leave the font set to Arial but change the size to 9.

ⓗ Set the color to Solar Yellow and click the Bold button **B**.

ⓘ Now, next to the Label Field setting, click Expression.

A label expression lets you add custom text to an attribute value to make a more informative label.

ⓙ In the Label Expression dialog box, highlight and delete the current expression.

ⓚ In the list of fields, double-click ACRES to add it to the expression box.

ⓛ Type the rest of the expression as follows: **& " Acres"**.

Check your expression against Figure 8-39.

The expression tells ArcMap to take the value from the ACRES field, then tack on a space and the word "Acres." Without the expression, the map reader wouldn't know what the units were.

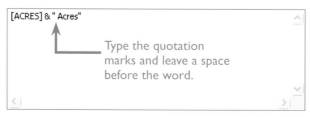

Figure 8-39

ⓜ Click OK on the Label Expression dialog box and the Layer Properties dialog box.

Ⓝ On the Layout toolbar, click the Zoom to 100% button 🔳.

The labeled feature should look similar to Figure 8-40.

Ⓞ On the Layout toolbar, click the Zoom Whole Page button 🔳.

With this label, Maplex has already given us a subtle placement advantage. Although we could create the same expression with the standard label engine, the text wouldn't be stacked. (If you want to see for yourself, open the data frame properties and temporarily switch back to the standard label engine.)

Figure 8-40

Since there isn't enough room to put all the inset maps right next to their sites, we'll move them to convenient locations, as shown in the layout sketch. Later, we'll connect them to the sites with leader lines.

Ⓟ On the map, make sure the Site 1 data frame is selected. (You should see the blue selection handles.)

▷ If necessary, click on the data frame to select it.

Ⓠ Drag the data frame to the lower right corner of the map as shown in Figure 8-41.

5) Create the rest of the inset maps. The rest of the inset maps will be the same size as Site 1 and will have the same layers with the same symbology. Instead of inserting new data frames, we'll just copy and paste the one we have, then zoom it to the appropriate site.

Ⓐ On the map, right-click the Site 1 data frame and choose Copy from the context menu.

Ⓑ Right-click anywhere on the main map and choose Paste.

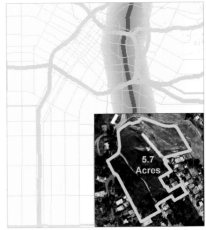

Figure 8-41

A new data frame, identical to Site 1 (and with the same name), appears on the layout and in the Table of Contents.

Ⓒ In the Table of Contents, click on the name of the active Site 1 data frame (the new, boldfaced one) to highlight it. Click again to make its name editable.

Ⓓ Rename the data frame to **Site 2** and press Enter (Figure 8-42).

Figure 8-42

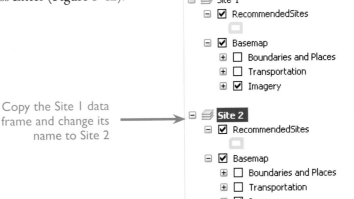

Copy the Site 1 data frame and change its name to Site 2

Figure 8-43

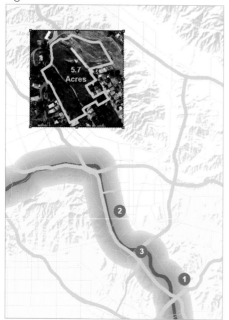

Site 2 is not right next to Site 1. The SiteID numbering follows the OBJECTID numbers assigned by ArcMap.

Figure 8-44

E On the layout, drag the Site 2 data frame to the approximate position shown in Figure 8-43.

The Site 2 inset map is still zoomed in on the Site 1 feature.

F In the Table of Contents, in the Site 2 data frame, open the attribute table for *RecommendedSites*.

G If necessary, move the attribute table away from the Site 2 data frame.

H In the attribute table, right-click the small gray box next to the second record (SITEID 2) and choose Zoom To.

I Close the attribute table.

The inset map zooms in to Site 2 (Figure 8-44). In each inset map, we're letting the site boundary fill the data frame. That means the inset maps will be at different scales from one another—a fact we'll note on the map later.

J Repeat the process to create three more inset maps. Use Figure 8-45 to guide your placement of the data frames. In summary, the steps are:

- Copy and paste an existing data frame.
- Rename the new data frame.
- Drag it to the appropriate place on the layout.
- Open the *Recommended Sites* attribute table and zoom to the feature with the matching SITEID number.

K When you're finished, save the map.

L If necessary, in the Table of Contents, drag and drop the inset map data frames into sequential order.

  ▷ Optionally, collapse them in the Table of Contents.

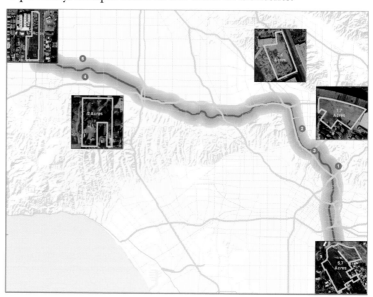

Figure 8-45

6) Set label placement properties. Four of the site labels are well-placed, but the one at Site 5 is squeezed against the edges of the polygon boundary. It will look better if we turn it sideways.

Ⓐ If necessary, make Site 5 the active data frame.

▷ How? Right-click the data frame name in the Table of Contents and choose Activate.

Ⓑ In the Table of Contents, in the Site 5 data frame, open the layer properties for *RecommendedSites*. Click the Labels tab, if necessary.

Ⓒ In the lower left corner of the dialog box, click Placement Properties.

Ⓓ In the Placement Properties dialog box, make sure the Label Position tab is selected and the drop-down list is set to Regular Placement.

Ⓔ Click the Position button. In the Position Options dialog box, click Straight (Figure 8-46).

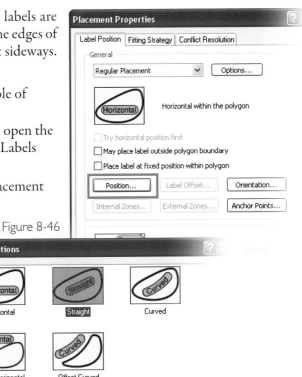

Figure 8-46

The straight setting orients the label to the longest axis in the polygon feature.

Ⓕ Click OK on all open dialog boxes.

The label for Site 5 is rotated to fit the feature better (Figure 8-47). Depending on your system and settings, some of your labels may look slightly different (for example, the text may not always be stacked).

Figure 8-47

7) Create a graph of median household income. We know each site meets our minimum demographic requirements, but it would be nice to compare them graphically to find out if one or more stand out as especially well-qualified. In this step, we'll create a bar graph of median household income. Later, we'll add preexisting graphs to show population density and the percentage of children.

Ⓐ Activate the Main Map data frame.

Ⓑ From the main menu, choose View ➔ Graphs ➔ Create to open the Create Graph Wizard.

Ⓒ At the top of the dialog box, click the Graph Type drop-down arrow and choose Horizontal Bar (under the Bar folder).

Ⓓ Click the Layer/Table drop-down arrow and choose *Recommended Sites*.

Ⓔ Set the Value field to Median Household Income.

Ⓕ Skip the Y field (optional) setting.

Ⓖ Set the Y label field to SiteID.

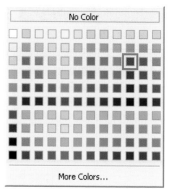

Figure 8-48

In the preview window, the graph is starting to take shape. The five sites are listed on the y-axis and incomes are shown on the x-axis.

H Uncheck the "Add to legend" box.

I Click the Color drop-down arrow and choose Custom.

J Click the color patch that appears next to this setting. On the color palette, choose Cretean Blue (Figure 8-48) or another color you think will look good on the map.

K Compare your graph preview and settings to Figure 8-49, then click Next.

There are six settings to make on this panel.

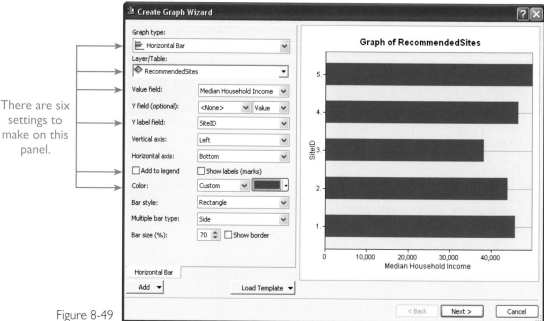

Figure 8-49

L In the next panel, in the Title box, replace the default title (Graph of Recommended Sites) with **Median Household Income** (Figure 8-50).

Figure 8-50

The title makes the x-axis label redundant. We'll address this in the next step.

M Click Finish.

The graph is displayed in a floating window that is not yet part of the map layout (Figure 8-51). You can move it outside the virtual page and even outside the ArcMap window (except that ArcMap should be maximized). The graph is also live-linked to the data: if you were to edit a value in the attribute table, the graph would update automatically.

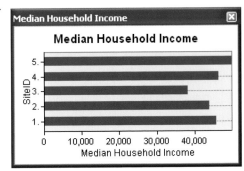

Figure 8-51

8) **Set advanced properties for the graph.** The three graphs are fairly small design elements in the layout. We'll make the axis labels a little bigger so they'll be legible on the final map.

🅐 Right-click anywhere on the graph window and choose Advanced Properties to open the Editing dialog box.

🅑 In the left-hand window, under Chart, expand Axis. Click Left Axis to highlight it.

In the window on the right, a set of tabs appears with all the possible settings for the left axis (that is, the y-axis).

🅒 Click the Labels tab (top row), then the Format tab (bottom row).

🅓 On the Format tab, click the Values Format drop-down arrow and choose 0 (Figure 8-52).

This will remove the decimal point that follows each of the SiteID numbers.

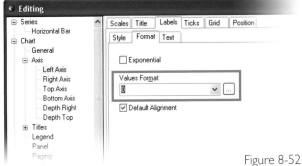

Figure 8-52

🅔 Click the Title tab (top row), then click the Format tab underneath it.

🅕 Click the Text tab (third row) and make sure the Font tab (fourth row) is selected.

🅖 Change the Size to 10 and check the Bold box (Figure 8-53).

The graph updates as you make changes, so it's easy to experiment.

🅗 In the left-hand window, click Bottom Axis.

🅘 Again, change the Size to 10 and check the Bold box.

🅙 Click the Style tab (second row).

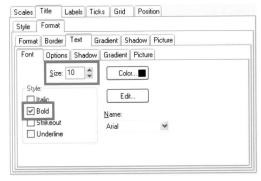

Figure 8-53

Figure 8-54

🅚 Replace the title (Median Household Income) with **Dollars ($)** (Figure 8-54).

🅛 On the Editing dialog box, click Close to accept your changes.

Now that the graph looks the way we want, we can make it an element of the layout.

Ⓜ Right-click anywhere on the graph and choose Add to Layout from the context menu.

The graph window remains open and the graph also appears as a selected element on the layout (Figure 8-55).

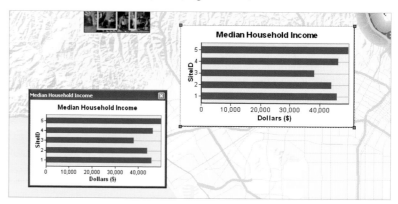

Figure 8-55

The graph is saved with the map document and can be accessed at any time from the View menu. (It can also be saved as a file to disk.) The layout element can now be moved, resized, and otherwise treated as a graphical object on the virtual page. It still maintains its connection to the graph. If you edit properties in the graph window, or edit data values in the *RecommendedSites* attribute table, these changes will be reflected in the layout element.

Ⓝ Close the Median Household Income graph window.

On the layout, the graph is too big.

Ⓞ Right-click the layout graph and choose Properties. On the Properties dialog box, click the Size and Position tab.

Ⓟ In the Size area, change the current Width to **1.75** and make sure the Preserve Aspect Ratio box is checked.

Ⓠ Check your settings against Figure 8-56, then click OK.

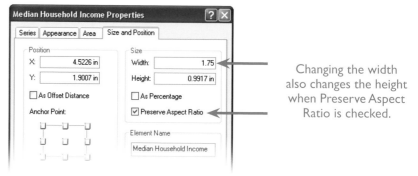

Changing the width also changes the height when Preserve Aspect Ratio is checked.

Figure 8-56

For now, you can just leave the layout graph where it is on the page. We'll add two more graphs in the next step and then move and position all the graphs together.

9) Add graphs of population density and percentage of children. The two remaining graphs have already been created and we'll just add them to the layout. (If you prefer to create them yourself from scratch, that's fine: follow the same steps you used to create the median household income graph and make appropriate changes for the value field, graph title, and bottom axis title.)

Ⓐ From the main menu, choose View → Graphs → Load.

Ⓑ In the Open dialog box, navigate to C:\UGIS\ ParkSite\MapsAndMore.

Ⓒ Click *Percent of Children.grf* to select it, then click Open to add the graph window to ArcMap (Figure 8-57).

Ⓓ Right-click the Percent of Children graph and choose Add to Layout, then close the graph window.

Ⓔ In the same way, load *Population Density.grf* (Figure 8-58).

Ⓕ Add the Population Density graph to the layout and close the graph window.

The layout graphs may all be on top of each other.

Ⓖ Drag the Population Density and Percentage of Children graphs to other parts of the layout so you can see all three graphs.

Ⓗ Click the Population Density graph to select it. Hold down the Shift key and click first on the Percentage of Children graph, then the Median Household Income graph.

All three graphs should be selected (Figure 8-59).

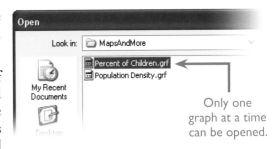

Only one graph at a time can be opened.

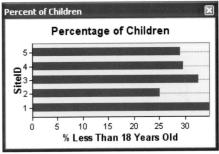

Figure 8-57

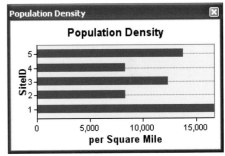

Figure 8-58

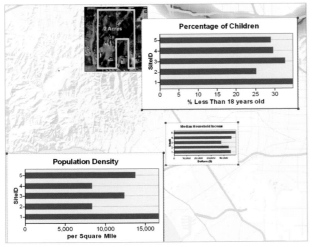

Figure 8-59

ⓘ Make sure that Median Household Income is the last selected element. (It should have blue selection handles; the other two graphs should have green handles.)

  ▷ The selection order affects the next step, so, if necessary, unselect the graphs and select them again.

Ⓙ Right-click any of the selected elements and choose Distribute → Make Same Size.

The two big graphs are resized to match the little one (the last element selected).

Ⓚ Click outside the layout page to unselect the elements.

10) Place the graphs. In the layout sketch, the graphs are arranged vertically on the left side of the layout. We'll align them just a tiny bit away from the edge of the layout using a guide.

Ⓐ At the top of the layout page, place the mouse pointer over the top ruler and click to create a guide.

Guides are nonprinting lines, light blue on the layout, that you can use to align and position elements. The current position of the guide is marked on the top ruler with a small gray arrow.

Ⓑ On the top ruler, click and drag the guide to the 0.30 inches mark.

As you drag the guide, its current position on the ruler is shown as a tip (Figure 8-60).

A tip shows the current position.

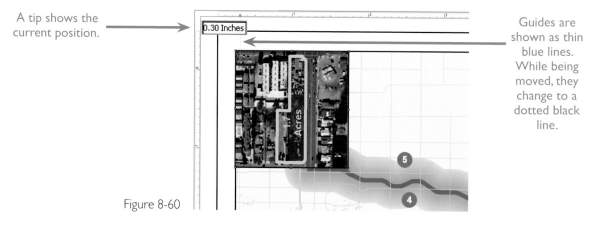

Guides are shown as thin blue lines. While being moved, they change to a dotted black line.

Figure 8-60

Ⓒ If necessary, click the Select Elements tool ▸ on the Tools toolbar. Click on the Site 5 inset map (in the upper-left corner of the layout) to select it.

Ⓓ Drag the inset map away from the guide and then back so that it snaps to the guide.

Ⓔ Click and drag the three graphs, one at a time, until they snap to the guide.

F Arrange them more or less as shown in Figure 8-61.

G Hold the Shift key and select all three graphs (in any order).

H Right-click any selected graph and choose Distribute → Distribute Vertically.

The vertical spacing between the graphs is made equal.

I Right-click again on a graph and choose Group.

J Click outside the layout page to unselect the grouped elements.

In the next exercise, we'll label some features, add a title, descriptive text, and some common map elements like a scale bar and north arrow. At this point, the layout should look like Figure 8-62.

Figure 8-61

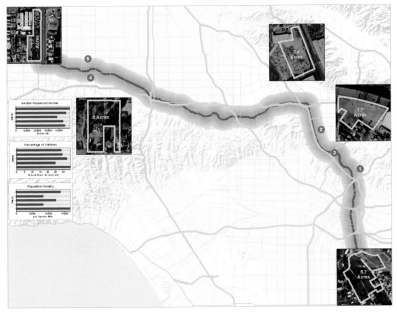

It doesn't matter what the top-to-bottom order of the graphs is.

Sites 1 and 3 come out best in a purely demographic comparison.

Figure 8-62

K Save the map.

L If you're continuing to the next exercise now, leave ArcMap open; otherwise, exit ArcMap.

Managing graphs

Graphs in a map document can be opened at any time from the Graphs command on the View menu. Open the Graph Manager (View > Graphs > Manage) to list all graphs and access their properties. Graphs can be saved as files with the extension .grf and loaded into other ArcMap documents. Graphs can also be exported as images in standard file formats. A graph references a layer or table in the map document in which it is created. That layer or table in turn references a dataset on disk. When a graph created in one map is saved and loaded into another, the second map needs to have the same layer, with the same value field, that the graph was originally based on.

Exercise 8c: Finish the map

The map still needs labels to identify prominent places and major roads. We'll label these features, then convert the labels to a special kind of text called annotation that will give us complete manual control over placement. We also need to draw leader lines connecting the inset maps to their corresponding sites on the main map. Finally, we'll add a title, scale bar, north arrow, and acknowledgments.

1) Add and symbolize cities. To put some recognizable place names on the map, we'll add and label the *Cities* layer while making the features themselves invisible.

Ⓐ If necessary, start ArcMap and open Lesson8 from the list of recent maps.

Ⓑ In the Table of Contents, activate the Main Map data frame.

Ⓒ Open the data frame properties for Main Map and click the General tab, if necessary.

Ⓓ Near the bottom of the dialog box, set the Label Engine to ESRI Maplex Label Engine, then click OK.

Ⓔ From the Catalog window, under ReadyData.gdb, add *Cities* to the layout.

In the Table of Contents, the layer should appear directly under the *LARiver* layer, which is where we want it.

Ⓕ In the Table of Contents, rename the *Cities* layer to `City Labels`.

▷ How? Click the layer name to highlight it. Click again to make it editable.

Ⓖ Open the Symbol Selector for the *City Labels* layer.

Ⓗ In the Symbol Selector, set the Fill Color to No Color and set the Outline Color to No Color. Click OK.

Now the city features are invisible on the map.

2) Label cities. So far, we've done all our labeling from the Layer Properties dialog box. We can also use the Labeling toolbar, which is more convenient in some ways—for one thing, it lets you label *all* layers in the map from the same dialog box.

Ⓐ Add the Labeling toolbar (Figure 8-63).

▷ How? From the main menu, choose Customize → Toolbars → Labeling.

Figure 8-63

B On the Labeling toolbar, click the Label Manager button 🖹.

In the Label Manager, each layer in the data frame is listed in the box on the left. A check mark indicates whether labels are turned on. (Right now, only *RecommendedSites* is labeled.) Underneath each layer name is a label "class" called Default. In the Default label class, all features in the layer are labeled the same way. If you wanted to vary the labeling properties within a layer (for example, to label large and small cities with different type sizes), you would create additional label classes.

Labeling properties for a layer are set on the right side of the Label Manager and with other tools on the Labeling toolbar. The functionality is the same whether you use the Labeling toolbar or the Labels tab in the layer properties—how you work with labels is really a matter of preference.

C In the Label Manager, check the box next to *City Labels*.

D Under *City Labels,* click its Default label class to highlight it.

The property settings on the right side of the Label Manager now apply to the *City Labels* layer.

E In the Text String area, make sure the Label Field is set to NAME.

F In the Text Symbol area, set the font size to 9 and make it bold.

G Set the color to Gray 40%.

H Check your settings against Figure 8-64, then click OK.

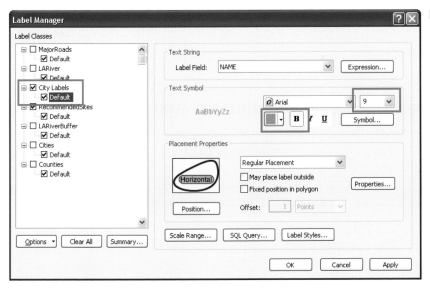

Figure 8-64

The city labels appear on the map. (The positions of your labels may not match Figure 8-65 exactly.) Even with Maplex at work, you're likely to notice some problematic placements, such as:

- Labels crossing freeways
- Labels partially covered by inset maps
- Labels too close to the edge of the layout

Figure 8-65

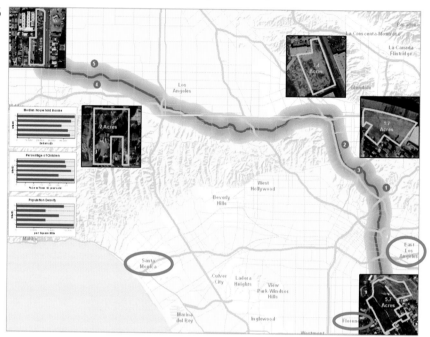

We also want to be selective about which labels are drawn. It's helpful to mark well-known places like Beverly Hills and Santa Monica, but maybe not places that are less familiar (Ladera Heights, Westmont) or peripheral to our map (Pasadena). We also want to make sure that the label for Los Angeles is centrally placed.

We could resolve some of the problems with adjustments to labeling rules, but dynamic labeling will never give us complete control of placement. At this point, it makes sense to convert the labels into a type of text called map annotation that we can place at will. (For more about labeling and annotation, see the topic *Labels, annotation, and graphic text* on page 327.)

3) **Convert labels to map annotation.** We'll convert the city labels to map annotation. This is text that is stored in the map document, managed as a data frame property, and manipulated with tools on the Draw toolbar.

Ⓐ In the Table of Contents, right-click *City Labels* and choose Convert Labels to Annotation.

B In the Convert Labels to Annotation dialog box, in the Store Annotation area, click "In the map."

You can store annotation as a geodatabase feature class, making it similar to point, line, or polygon feature classes. You would do this if you planned to reuse the annotation in other maps. In our case, the annotation serves the purposes of just this particular map, so it makes sense to store it in the map document.

C In the Create Annotation For area, click "Features in current extent."

D At the bottom of the dialog box, uncheck the box "Convert unplaced labels to unplaced annotation."

E Check your settings against Figure 8-66, then click Convert.

Unlike labels, which stay the same size at all map scales, annotation changes size relative to the reference scale.

Annotation will be stored in the map, not as a geodatabase feature class.

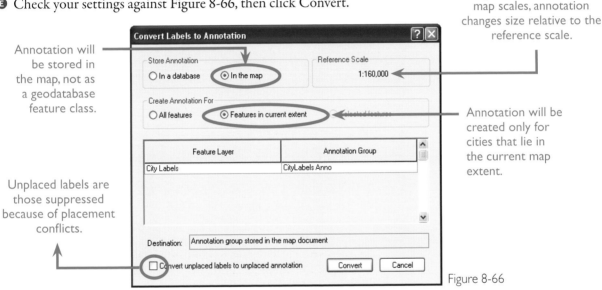

Unplaced labels are those suppressed because of placement conflicts.

Annotation will be created only for cities that lie in the current map extent.

Figure 8-66

Although the layout looks the same, something significant has happened behind the scenes. ArcMap placed annotation where the labels used to be and automatically turned off dynamic labeling.

F Open the Main Map data frame properties and click the Annotation Groups tab (Figure 8-67).

Every data frame has a <Default> annotation group, which holds text or graphics you create with the Draw toolbar in data view. (The black ellipses that marked your guesses in Lesson 1 were pieces of annotation stored in the <Default> group.) When you convert labels to annotation, a new group is created automatically. The annotation stored in different groups can be managed independently.

G Close the Data Frame Properties dialog box.

Figure 8-67

4) Edit the City Labels annotation group. We'll use the tools on the Draw toolbar to position or delete individual pieces of annotation.

Ⓐ Add the Draw toolbar (Figure 8-68) and confirm that the Select Elements tool ▶ is selected.

Figure 8-68

At the moment, you can't select pieces of annotation, only layout elements like the graphs or the data frames themselves. On a layout, two types of elements coexist: those that inhabit page space and those that inhabit map space. The locations of elements in page space are defined relative to the layout rulers. The locations of elements in map space—like our city annotation—are defined relative to a real-world coordinate system (in this case, State Plane California Zone 5). To work in map space, you have to "focus" the data frame.

Ⓑ On the Layout toolbar, click the Focus Data Frame button 🖾.

A thick, diagonal-line border appears around the data frame. We have left page space and entered map space.

Figure 8-69

Move annotation off the freeway

Ⓒ Click the Santa Monica piece of annotation to select it. Drag it away from the freeway, but keep it in the gray area that separates Santa Monica from Los Angeles (Figure 8-69).

▷ Here, and in the rest of the step, modify the instructions as needed if your annotation isn't placed the way it's shown in the figures.

The Los Angeles annotation is too far northwest: it's near the geographic center of the city, but not near downtown, which is probably a more meaningful location.

Ⓓ Click and drag the Los Angeles annotation toward downtown (think in terms of Dodger Stadium), as shown in Figure 8-70.

Some pieces of annotation are partially covered by inset maps. We'll delete these.

Figure 8-70

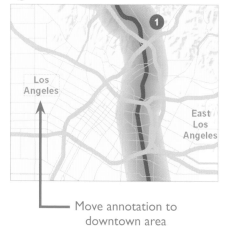

Move annotation to downtown area

Ⓔ Select the Florence annotation (lower right, behind the Site 1 inset map) and press the Delete key on your keyboard.

Ⓕ In the same way, delete the La Crescenta - Montrose annotation (upper right, behind the Site 2 inset map).

Ⓖ Delete any other annotation that is partially covered by an inset map.

There are actually some pieces of annotation that are completely hidden behind inset maps. We could temporarily move the inset maps to find and delete (or reposition) these, but since they're not showing, we'll just leave them where they are.

Some other annotation is superfluous, identifying cities that aren't landmarks or that lie too far from the area of interest. Deciding which of this annotation to delete is subjective. On our map, we've deleted the following pieces (again, your annotation may not include all the same cities):

- Pasadena
- La Canada - Flintridge
- Ladera Heights
- View Park - Windsor Hills
- Westmont
- Malibu

Ⓗ Delete or reposition any other annotation to suit your cartographic sense.

In our map, we moved East Los Angeles slightly north to get it off a freeway, and we moved Glendale southwest, a little farther away from the inset maps for Sites 2 and 3.

Ⓘ When you're finished, click in some neutral part of the data frame (like the ocean) to unselect the last piece of annotation.

Your map should look more or less like Figure 8-71:

Figure 8-71

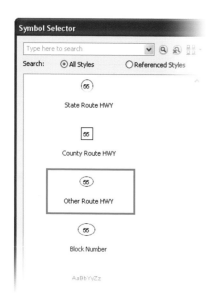

Figure 8-72

5) Place highway markers. We want to label the major roads, but not with their names, which would be hard to place legibly on our map. Instead, we'll use their numeric designations.

Ⓐ On the Labeling toolbar, click the Label Manager button 🏠.

Ⓑ In the Label Manager, check the box next to *MajorRoads*.

Ⓒ Under *MajorRoads,* click its Default label class to highlight it.

The property settings on the right now apply to the *MajorRoads* layer.

Ⓓ In the Text String area, set the Label Field to HWY_SYMBOL.

This field holds numeric descriptors for roads that have them.

Ⓔ In the Text Symbol area, click Symbol.

The Symbol Selector lists predefined text symbols, just like the predefined point, line, and polygon symbols you've seen earlier.

Ⓕ Scroll down and locate the Other Route HWY symbol near the bottom (Figure 8-72).

Ⓖ Double-click the symbol to select it and automatically close the Symbol Selector.

Ⓗ In the Placement Properties area, confirm that Regular Placement is selected.

Ⓘ Click Position. In the Position Options dialog box, click Centered Horizontal and click OK.

Ⓙ Click Properties. In the Placement Properties dialog box, click the Conflict Resolution tab.

Ⓚ Check the Remove Duplicates box, then click Limits.

Ⓛ In the Duplicate Labels dialog box, for the Search radius, type **4**. Set the units to Inches (Figure 8-73), then click OK on all open dialog boxes.

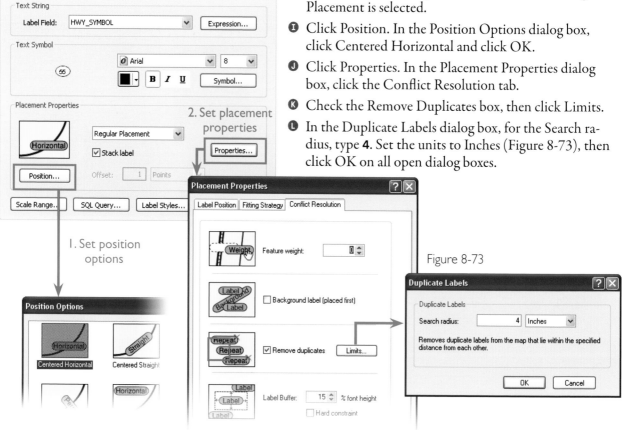

Figure 8-73

On the map, the major roads are labeled with text symbols (Figure 8-74). Duplicates are removed unless they're more than four inches apart. We could make slight placement improvements by converting the labels to annotation, but things are pretty good as they are.

6) **Add leader lines.** We placed the inset maps away from the site locations, so we need to connect them to their red symbols with leader lines.

Ⓐ If the data frame is still in focus, on the Layout toolbar, click the Focus Data Frame button 🖼 to remove the focus.

The diagonal-line border should be gone.

Ⓑ On the Layout toolbar, click the Zoom In tool. Zoom in on the Site 5 inset map and its corresponding symbol in the upper left corner.

Ⓒ On the Draw toolbar, click the Drawing drop-down menu and choose Default Symbol Properties.

Ⓓ In the Default Symbol Properties dialog box, click Line to open the Symbol Selector.

Ⓔ Set the Color to Mars Red and the Width to **0.5**.

Ⓕ Click OK on the Symbol Selector. Compare your settings to Figure 8-75, then click OK on the Default Symbol Properties dialog box.

Ⓖ On the Draw toolbar, click the drop-down arrow next to the Rectangle tool and choose Line (Figure 8-76).

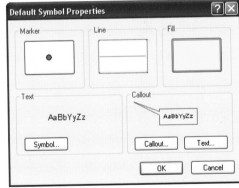

Figure 8-74

Two of the highway symbols are a bit close to the Site 2 inset map, but it might be simpler to move the data frame than to convert the labels to annotation.

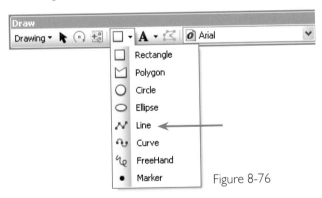

Figure 8-76

Figure 8-75

H Click on top of the Site 5 symbol to start the line. Go straight up, then click to go left. Double-click on the edge of the inset map to complete the line (Figure 8-77).

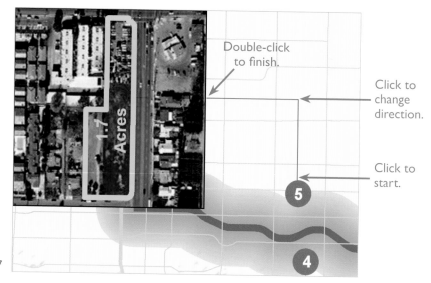

Figure 8-77

Text or graphics you create with drawing tools when the data frame is *not* in focus—like this leader line—are in page space and belong only to the layout. These elements won't appear in data view.

I On the Layout toolbar, click the Zoom Whole Page button.

J Click the Zoom In tool and zoom in on another inset map, such as Site 4, and its corresponding symbol.

K On the Draw toolbar, click the Line tool and draw a leader line connecting the symbol to the inset map.

Not all the leader lines need a right angle; some can be connected with a straight line. It's mostly a matter of what you think looks good. Refer to the finished map on page 333 for ideas.

L Add leader lines for the remaining sites in the same way.

  ▷ If a line turns out badly, delete it and draw another one.

M When you're finished, click the Zoom Whole Page button and unselect the last leader line.

N Save the map.

7) **Add the map title**. We want the title to have the largest and most prominent typeface on the layout.

A From the main menu, choose Insert → Title to open the Insert Title box.

The title that we entered earlier in the map document properties appears automatically at the top of the layout.

B Right-click the selected title and choose Properties.

In the Properties dialog box, on the Text tab, the title appears as a dynamic text tag (Figure 8-78).

Figure 8-78

C Click Change Symbol.

D In the Symbol Selector, make the Size 22 and the Style Bold. Click OK.

E Click the Size and Position tab (Figure 8-79).

Figure 8-79

The X,Y position values are in inches, because the map title is graphic text that belongs to page space.

Dynamic text updates automatically when you change the property it's based on.

F Click OK on the Properties dialog box.

G Drag the title to the lower left corner of the layout.

The position of the title is specified in the layout's ruler units.

This corner of ocean is empty space that we can use for the map title, north arrow, and scale bar. The title is too long, however. To put it on two lines, we have to convert it to graphics, which will break the dynamic text link.

H Right-click the selected title and choose Convert to Graphics.

I Right-click the title again and choose Properties. In the Properties dialog box, click the Text tab.

J In the Text box, put the cursor between the words "River" and "Park." Delete the space between them, then press the Enter key (Figure 8-80). Click OK.

Figure 8-80

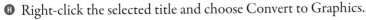

Text:

Potential LA River
Park Site Locations

K Snap the text to the guide and place it over the ocean as shown in Figure 8-81.

L Click in an empty part of the ocean to unselect the text.

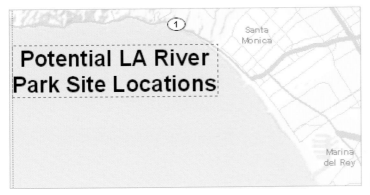

Figure 8-81

Figure 8-82

Draw splined text along this path with a few clicks.

Figure 8-83

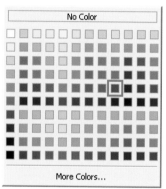

Figure 8-84

If you switch to data view, you'll see the Pacific Ocean annotation, but not the map title.

8) Add Pacific Ocean annotation. Speaking of the ocean, we should identify it as the Pacific on the map. We don't have an ocean feature to label (the ocean is part of the *Terrain* basemap), so we need to create a piece of text. This text rightfully belongs to the map body, so we'll add it as map annotation.

Ⓐ On the Draw toolbar, click the Drawing menu, choose Active Annotation Target, and confirm that <Default> is checked.

We don't want to add this annotation to the CityLabels Anno group.

Ⓑ On the Layout toolbar, click the Focus Data Frame button 📱 to put the data frame in focus.

Ⓒ On the Draw toolbar, click the New Text **A** drop-down arrow and choose Splined Text 🝆.

With the Splined Text tool you define a curved path for your text before you type it. We'll draw a gentle curve more or less following the coastline between Santa Monica and Marina del Rey.

Ⓓ Click once to begin the text path, click again to define a midpoint, then double-click to finish the path (Figure 8-82).

When you double-click, an editable text box opens.

Ⓔ In the text box, type **Pacific Ocean** and press Enter.

The text is drawn on the curved path you typed.

Ⓕ Double-click the selected annotation to open its properties.

Ⓖ On the Text tab, click Change Symbol.

Ⓗ In the Symbol Selector, make the color Delft Blue (Figure 8-83).

Ⓘ Click the Font drop-down arrow and choose Palatino Linotype.

▷ If you don't have this font, leave the font set to Arial or choose something else.

Ⓙ Click the Size drop-down and choose 11.

Ⓚ Make the style Bold **B** and Italic *I* , then click OK in the Symbol Selector.

By convention, text for bodies of water is italicized.

Ⓛ Click the Size and Position tab.

The X,Y Position values are in feet. This annotation belongs to map space—its position is defined relative to the origin of the California State Plane Zone 5 coordinate system.

Ⓜ Click OK on the Properties dialog box.

Ⓝ Drag the text to the approximate position shown in Figure 8-84.

Ⓞ Click in an empty part of the ocean to unselect the text.

Labels, annotation, and graphic text

Labels are text associated with map features. Labels are a layer property you can turn on or off; the label text is drawn from a field in the layer attribute table. Labels are dynamically placed by ArcMap, subject to preferences you set, and are repositioned as you zoom or pan the map. ArcMap has both a standard label engine and a Maplex engine with advanced placement capabilities.

Cities labeled with Maplex

OBJECTID *	Shape *	NAME	
1	Point	Avalon	City
2	Point	Attu Station	Census De
3	Point	Anahola	Census De
4	Point	Fleele	Census De

Labels come from this field

Annotation is map text or graphics that you can place exactly where you want. It can describe map features or places on the map not represented by features: in general, it can be any text or graphics that belong to the body of the map. Annotation can be created piece-by-piece as you need it or by a batch conversion of labels. Converting labels to annotation is a common step in the later stages of map creation when you're ready to make final adjustments to text placement.

Annotation can be stored either in a map document or as a geodatabase feature class. How best to store it is mostly a question of reusability. If you plan to reuse the annotation, store it in a geodatabase; if not, store it in the map. Geodatabase annotation, like other feature classes, is edited in an edit session. Map annotation is edited with tools on the Draw toolbar.

Labels in the upper left map were converted to geodatabase annotation. Text strings and placement were then edited with annotation editing tools.

The *CitiesAnno* feature class can be added to any map document.

Graphic text includes the title, acknowledgments, and any marginal text that isn't part of the map body. It can only be added and seen in layout view. It can be placed outside or on top of a data frame, but either way its position is defined in layout page units, not real-world coordinates. Graphic text is added with tools on the Draw toolbar. It can also be inserted as dynamic text which updates automatically.

The Pacific Ocean text is stored as map annotation in the <Default> group. The LA Airport text and symbol are stored in the Airports group.

The date is dynamic graphic text. The credits are static graphic text.

Figure 8-85

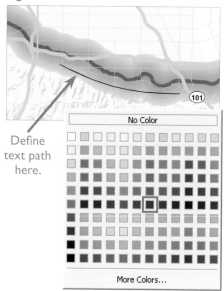

Define text path here.

Figure 8-86

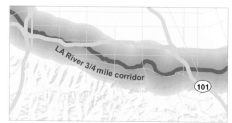

Figure 8-87

9) Add river buffer annotation. We'll add one more piece of annotation to clarify what the LA River buffer represents.

Ⓐ Make sure the data frame is still in focus.

Ⓑ On the Draw toolbar, click the Splined Text tool.

Ⓒ On the layout, define a text path inside the river buffer, south of the river, between the 405 and 101 freeways (Figure 8-85).

▷ If you want, you can click more than once along the path.

Ⓓ In the text box, type **LA River 3/4 mile corridor** and press Enter.

Ⓔ Double-click the selected annotation to open its properties.

Ⓕ On the Text tab, click Change Symbol.

Ⓖ In the Symbol Selector, make the color Fir Green (Figure 8-86).

Ⓗ Leave the font set to Arial, but change the Size to 8.

Ⓘ Make the style Bold **B**. Click OK in the Symbol Selector and the Properties dialog box.

Ⓙ Drag the text to the approximate position shown in Figure 8-87.

Ⓚ Click somewhere else in the data frame to unselect the text.

Ⓛ On the Layout toolbar, click the Focus Data Frame button to take the data frame out of focus.

10) Add a scale bar to the main map. A scale bar will give the map reader a sense of distance on the map.

Ⓐ From the main menu, choose Insert → Scale Bar to open the Scale Bar Selector.

In the box on the left, you can choose from a variety of scale bar types.

Ⓑ Click Single Division Scale Bar to select it (Figure 8-88).

This is a simple, easy-to-read scale bar.

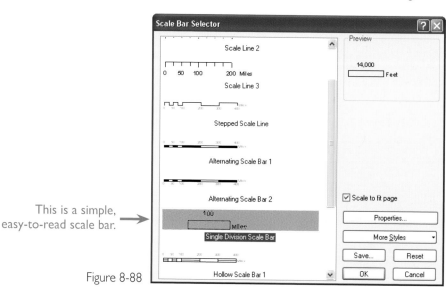

Figure 8-88

C Click Properties.

D In the Scale Bar Properties dialog box, make sure the Scale and Units tab is selected.

E In the Units area, set the Division Units to Miles. Accept the other default settings on this tab (Figure 8-89).

F Click the Format Tab.

G In the Text area, change the Size to 9.

H In the Bar area, click Symbol to open the Symbol Selector.

I Set the Fill Color to Arctic White and the Outline Width to **0.5**.

J Click OK on all open dialog boxes.

The scale bar draws on the map, but it's longer than we want.

K Make the scale bar a little bit shorter by dragging one of its side selection handles.

When you release the mouse button, the distance value is adjusted automatically.

L Resize the scale bar until its value is 3 Miles.

M Drag the scale bar to the bottom of the layout and place it approximately as shown in Figure 8-90.

N Unselect the scale bar.

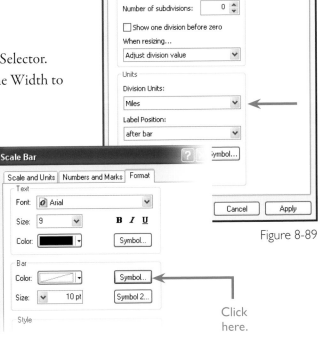

Click here.

Figure 8-89

Scale bars and scale text

Scale bars provide a visual indication of the size of features and distance between features on the map. A scale bar is a line or bar divided into parts and labeled with its ground length, usually in multiples of map units such as tens of kilometers or hundreds of miles. If the map is enlarged or reduced, the scale bar remains correct. You can also represent the scale of your map with scale text, which is a verbal expression of scale. Scale text may relate equivalent units ("one inch equals 2 miles") or be a representative fraction that holds for any units ("1:100,000"). Scale text will be wrong if the map is enlarged or reduced after it has been printed.

11) Add a north arrow. We'll add a north arrow to remove any ambiguity about directions.

A From the main menu, choose Insert → North Arrow.

B In the North Arrow Selector, scroll down and click ESRI North 19 to select it.

We picked this ornamental compass rose as a matter of taste—feel free to choose any north arrow you like.

Figure 8-90

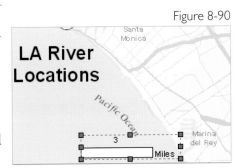

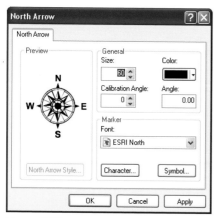

Figure 8-91

C Click Properties.

D In the North Arrow dialog box, set the Size to 60 (Figure 8-91).

E Click OK on both open dialog boxes.

F Place the north arrow approximately as shown in Figure 8-92.

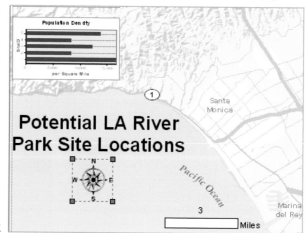

Figure 8-92

G Unselect the north arrow.

12) Add data attribution. It's important to include data sources, attribution, and any map reading instructions.

A On the Draw toolbar, click the Font drop-down arrow and choose Arial Narrow.

B Click the Size drop-down arrow and choose 7.

C Click the Text Color drop-down arrow **A** and choose Gray 50%.

D On the Layout toolbar, click the Zoom In tool and zoom in on the lower left corner of the map.

> You may want to set your view scale to about 200% so that you can clearly see what you're typing.

E On the Draw toolbar, click the drop-down arrow next to the Splined Text button and choose New Text **A** . Click in the lower left corner of the map to create an editable text box.

F Press Enter immediately, so the text box contains the word "Text."

G Double-click the selected text to open its properties.

H In the Properties dialog box, on the Text tab, replace the existing text with **Data Sources: City of Los Angeles Department of Public Works - Bureau of Engineering; Esri; TeleAtlas**.

I Put the cursor between the words "Department" and "of." Delete the space between them and press Enter (Figure 8-93). Click OK.

Figure 8-93

J On the map, drag the selected text almost to the bottom of the layout and snap it to the guide.

K On the Draw toolbar, click the New Text button **A** and click to create another text box in the same general area.

L In the text box, type: **Inset maps are at different scales.** Press Enter.

M Position the two pieces of graphic text approximately as shown in Figure 8-94.

Figure 8-94

N Click outside the layout page to unselect the text.

O On the Layout toolbar, click the Zoom Whole Page button.

P Save the map.

In the next step, we'll preview and print the map. If you see any final adjustments you want to make, now is the time to do it. Look at the final map on page 333 for reference.

13) Preview and print the map. If you have a printer available, you can print your map.

A From the main menu, choose File → Page and Print Setup.

B At the top of the Page and Print Setup dialog box, accept the default printer name or choose a different one.

Our map layout has a horizontal orientation, so we need the paper to match.

C In the Paper area, click Landscape.

D In the Map Page Size area, check the "Use Printer Paper Settings" box.

E Check your settings against Figure 8-95, then click OK.

Figure 8-95

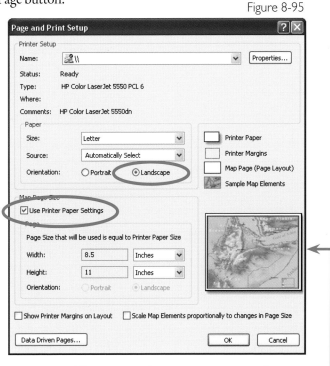

The paper and map page are correctly oriented with respect to each other.

With the map size matched to the printer paper, a drop shadow is added to the layout. Your layout is now a preview of what you'll see when you print.

F If you see any problems, fix them.

G Save the map.

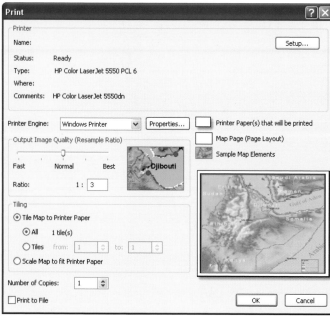

Figure 8-96

H On the Standard toolbar, click the Print button 🖨. On the Print dialog box, accept the defaults (Figure 8-96), then click OK to print your map.

I Optionally, remove the Draw, Labeling, and Layout toolbars from the interface.

J Exit ArcMap.

Hopefully, your map looks good. The next, and final, phase of the project is to share your analysis results online with a web map. You can see the lesson outline, and read a little bit about ArcGIS Online in the following pages; the lesson itself is a downloadable PDF file on the Understanding GIS Resource Center.

Our goal in this book has been to guide you through the main steps of a GIS analysis problem: to explore the study area; to state the problem in quantitative terms; to choose, prepare, and edit the data; to plan and carry out the analysis; to automate the analysis with a model; and to present the analysis results. Not every GIS analysis problem is the same, but you should now understand the basic approach and have the requisite skills to take on the next project that comes your way.

Our second goal has been to introduce you to ArcGIS Desktop 10 software. In planning this book, we made a decision to use the software only insofar as it served the needs of the project. We tried to make the project complex enough that we would have reason to use a lot of functionality; nevertheless, we definitely didn't see or do everything. Keep exploring the software on your own until the trial license period expires, and visit the book's resource center at resources.arcgis.com/Understanding-GIS.

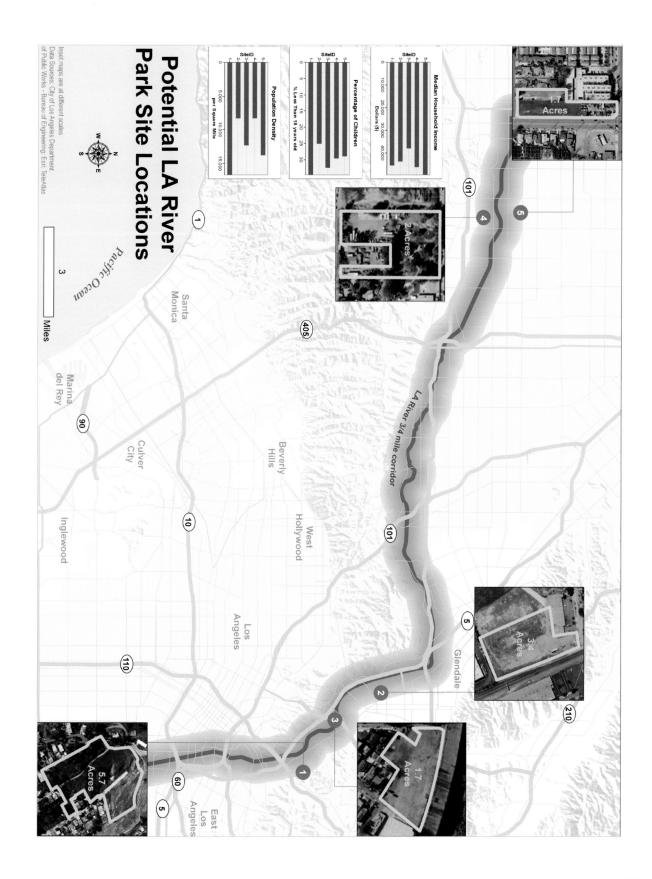

Potential LA River
Park Site Locations

Population Density
per Square Mile

SiteID
0   5,000   10,000   15,000

Percentage of Children
% Less Than 18 years old

SiteID
0   5   10   15   20   25   30

Median Household Income
Dollars ($)

SiteID
0   10,000   20,000   30,000   40,000

Inset maps are at different scales
Data Sources: City of Los Angeles Department
of Public Works - Bureau of Engineering; Esri; TeleAtlas

Pacific Ocean

Santa
Monica

Marina
del Rey

Culver
City

Inglewood

Beverly
Hills

West
Hollywood

Los
Angeles

Glendale

East
Los
Angeles

LA River 3/4 mile corridor

1.7
Acres

2 Acres

3.4
Acres

1.7
Acres

5.7
Acres

0   3   Miles

333

9 Share results online

THE CITY COUNCIL WOULD LIKE TO

make the park analysis available to as wide an audience as possible. A paper map, like the one we made in the last lesson, is the traditional way to share results, but online maps can reach a lot more people. In this lesson, we'll look at several ways to share maps and data digitally, but our first focus will be on creating a map presentation in ArcGIS Explorer Online.

ArcGIS Explorer Online is a free, web-based mapping application. Anyone with an Internet connection can use it to make their own maps from basemaps and other online map content—ArcGIS Desktop software isn't required. These maps can be saved and shared with other users at the ArcGIS.com website.

ArcGIS Explorer Online allows you not only to create interactive maps, but also to make map presentations. A map presentation is a series of map views, essentially a guided tour, that the map user follows. Each view, or slide, can be annotated with a title, pop-up windows, and other information.

In the first two exercises, we'll map our analysis results in ArcGIS Explorer Online, add narrative context with notes and pop-up windows, and create a map presentation. In the third exercise, we'll create a map package, which is a convenient way to share maps and data with other ArcGIS Desktop 10 users, and upload it to the ArcGIS.com website. In the last exercise, we'll go to ArcGIS.com, then test the map package by opening it in ArcMap.

Because the exercises in this lesson rely on online applications and websites that may change over time, this lesson is a PDF file, downloadable from the Understanding GIS Resource Center at resources.arcgis.com/Understanding-GIS. The next two pages offer the lesson outline and a quick overview of ArcGIS Online and its focal website ArcGIS.com.

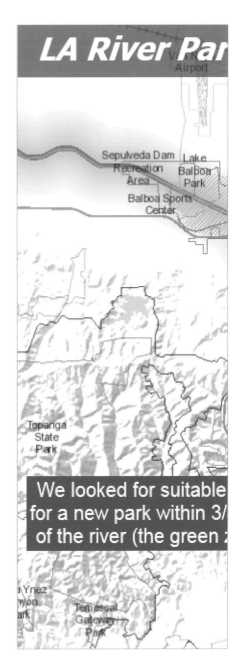

Lesson Nine roadmap

You are here:

What you'll do in this lesson:

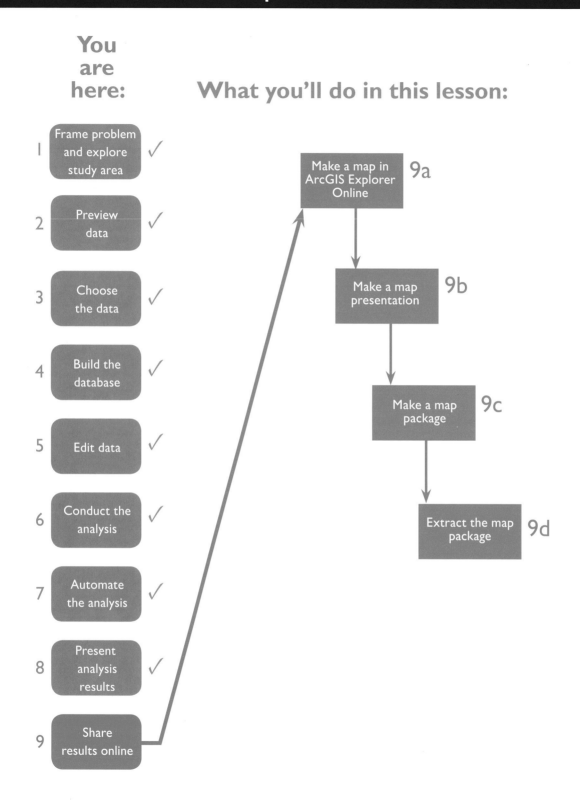

1 Frame problem and explore study area ✓

2 Preview data ✓

3 Choose the data ✓

4 Build the database ✓

5 Edit data ✓

6 Conduct the analysis ✓

7 Automate the analysis ✓

8 Present analysis results ✓

9 Share results online

Make a map in ArcGIS Explorer Online **9a**

Make a map presentation **9b**

Make a map package **9c**

Extract the map package **9d**

ArcGIS Online and ArcGIS.com

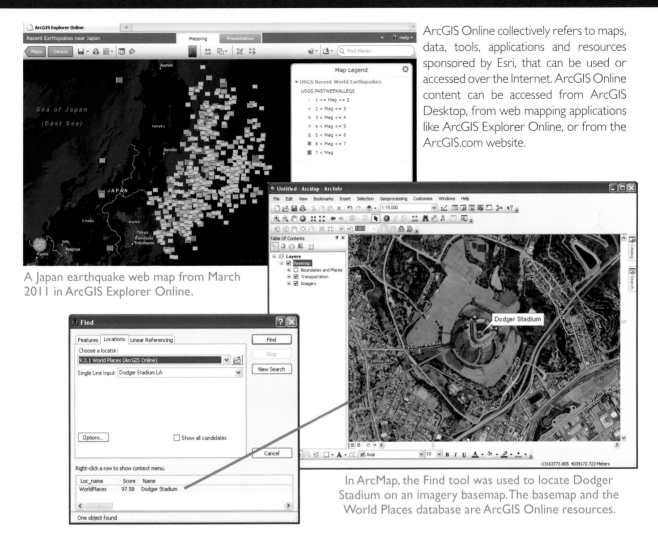

ArcGIS Online collectively refers to maps, data, tools, applications and resources sponsored by Esri, that can be used or accessed over the Internet. ArcGIS Online content can be accessed from ArcGIS Desktop, from web mapping applications like ArcGIS Explorer Online, or from the ArcGIS.com website.

A Japan earthquake web map from March 2011 in ArcGIS Explorer Online.

In ArcMap, the Find tool was used to locate Dodger Stadium on an imagery basemap. The basemap and the World Places database are ArcGIS Online resources.

ArcGIS.com is a website for using and managing ArcGIS Online resources. It has a web mapping application for viewing and making online maps. It offers access to custom map applications, built by Esri or members of the ArcGIS community, that run on the web and mobile devices. You can store and share content such as online maps you've made or map and layer packages you've uploaded from ArcGIS Desktop.

You can find online maps as well as mapping applications for the web and mobile devices in the ArcGIS.com gallery.

A Analysis issues

ONE THING YOU LEARNED FROM

the analysis is that the results you got—namely, the five sites recommended to the city council—depended on many different factors: the specific definitions you gave to the park siting guidelines in Lesson 2; the datasets you chose in Lesson 3; and several decisions you made in the course of data preparation and analysis. This is true for any analysis project: outcomes are determined by inputs, assumptions, and choices that are subject to reevaluation.

The scope and complexity of this project were somewhat limited by its context: after all, this is an introductory book about GIS analysis. In this appendix, we'll discuss a few issues generally relevant to analysis problems as well as some ways in which this analysis might be refined or expanded. We'll also touch on a few of its methodological soft spots. No GIS analysis is perfect or perfectly complete: being aware of limitations, and candid about them, is a strength.

General considerations

Data quality

Good results require good data. Ideally, data should be accurate to a known standard, complete, up-to-date, and supported by metadata. We worked with published data from commercial vendors and operational data maintained by a large municipal agency: you can't do much better. Still, we saw concerns to address. For example, in Lesson 4 we loaded a couple of features that were missing from our parks dataset. In Lesson 5 we edited a feature with an inaccurate boundary and added a missing feature. Issues like these may not reflect negatively on the data. For one thing, large, comprehensive datasets, like our parks dataset, are updated and published on schedules; they don't respond instantly to every change in the real

Top: a raster surface is composed of cells (pixels). Bottom: a TIN surface is composed of triangles.

world. All our demographic data, for example, was either published data from the 2000 census or demographic forecasts based on this data. When 2010 census data is published, the block group values will be different.

Also, data is generalized for use at particular scales. Our parks dataset was intended for use at scales of about 1:100,000. If you use data at a scale much larger or smaller than the scale it was designed for—or if your map mixes datasets that are meant for use at quite different scales—you're liable to encounter problems. Before criticizing the data, ask yourself if you're blaming a dog for not being a cat.

Data models

Our project relied on vector datasets (point, line, and polygon feature classes) for analysis. We used raster layers, in the form of basemaps, only for background display and context. In ArcGIS Desktop, a variety of geographic surfaces can be represented as raster datasets and analyzed with the ArcGIS extension products Spatial Analyst and 3D Analyst. These surfaces include elevation and its derivatives (slope, visibility), as well as soil, vegetation, temperature, rainfall, and many others. Another surface data model in ArcGIS is the triangulated irregular network (TIN), which represents surfaces as an expanse of contiguous triangles of varying size and shape (Figure A-1).

Had we incorporated a raster elevation surface and Spatial Analyst tools into our analysis, we might have considered slope as a park suitability factor. We could also have investigated questions like whether or not the river is visible from a particular site.

Figure A-2

The Modifiable Areal Unit Problem (MAUP)

The Modifiable Areal Unit Problem affects geographic analysis in general. The problem is that statistical data varies with the nature of the areal unit by which it is organized. An example from our analysis is shown in Figure A-2.

The triangular vacant parcel circled in yellow was rejected as a park site. Because it lies in a block group (pink striped area) with almost no population, it failed the population density requirement. However,

The population density of the block group (pink striped area) is just 16 people per square mile, while the census tract that contains it (orange polygon) has a density of 9,491 people per square mile. Our requirement was 8,000 people per square mile.

the census tract that contains the parcel *does* meet the density threshold. Had we used census tracts rather than block groups as our neighborhood unit—a valid choice—this parcel might have become a recommended site. (It might still have been disqualified on other grounds.) You can't really escape the Modifiable Areal Unit Problem. Had we used census tracts as our neighborhood units, we would surely have disqualified some parcels that satisfied requirements within their block group but not within the census tract.

Figure A-3

Distance analysis

Our analysis of distances was based on straight-line measurements, an oversimplification because people don't walk or drive in straight lines. Figure A-3 shows the situation with Recommended Site 1. The straight-line distance from the site to the river, represented by the dotted yellow line, is about 0.6 miles, putting the site inside the 0.75 mile limit. But the street distance to the nearest river access point is 1.6 miles, as shown by the green line and expressed in the Find Route dialog box (Figure A-4). The value of putting a park close to the river may not be defined entirely in terms of accessibility (for example, there might be value in seeing the river from the site), but accessibility is certainly a big part of it.

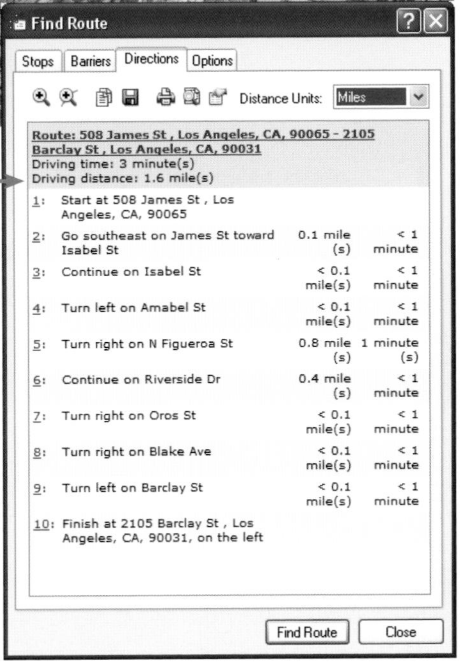

A sophisticated analysis of distance involves not only streets but also barriers. Even if the dotted yellow line in the figure followed a street, it wouldn't continue through the railroad yard on the river's eastern bank. Our analysis didn't account for barriers like railroad yards or freeways that complicate or block access. (Likewise, it didn't account for shortcuts and other local knowledge people have about navigating in their neighborhood.)

Figure A-4

To do complete network analysis, you need the Network Analyst extension for ArcGIS. This would allow you to create, for example, "service area" polygons showing the total street-accessible area around a site, either in terms of distance or travel time. Route-finding, as in the example on the previous page, can be done in ArcMap with no extensions.

Finding routes

On the Tools toolbar, click the Find Route button 🖫 to open the Find Route dialog box. On the Stops tab, you can add start and destination points by clicking on a map (any basemap, for example), by typing addresses, or by specifying a point layer in the map document.

Our use of straight-line distance has certainly understated true travel distances. Consequently, most of our recommended sites aren't as river-accessible as their distance values suggest. By the same token, the 0.25 mile value we used as a minimum distance from existing parks probably represents more than that in most travel scenarios.

Intended park use

One consideration missing from our guidelines was the intended use of the park. Although it's obviously a neighborhood-scale park, there are still distinctions to make: for example, is the park meant to support passive recreation (walking, picnicking, casual play) or organized sports and activities? This question leads to others about the need for facilities, parking, street access, landscaping, and so on; these questions, in turn, raise general GIS questions about slope and park shape. A hilly site or an oddly shaped one (a string bean, for example) might meet our other requirements but be unsuitable from the point of view of intended use.

A quantitative analysis of slope probably requires ArcGIS Spatial Analyst (see the section on *Data models*, above). Analyzing shapes to see whether one polygon fits inside another —for example, whether a baseball field will fit in a parcel—is a manual process that has to be evaluated with drawing, editing, and measurement tools. You can, however, use the Minimum Bounding Geometry tool to make a calculation of a shape's "regularity." The tool creates a feature class of minimum bounding shapes containing the input features (Figure A-5). (You can choose from various bounding shape options.) Dividing the area of the input feature by the area of the bounding geometry feature gives a regularity index—the closer the quotient is to 1, the more regular the shape.

Figure A-5

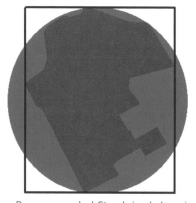

Recommended Site 1 (red shape) is shown against two different Minimum Bounding Geometry outputs: a feature envelope (blue outline) and a circle (gray).

Geoprocessing and spatial query on vacant parcels

Dissolving parcels

One of our guidelines was a minimum parcel size of one acre. In Lesson 4 we dissolved vacant parcels so we could consider the combined size of adjacent parcels that were individually smaller than an acre. This was useful for finding candidates (our five recommended sites are all composed of dissolved parcels), but it has some potential drawbacks.

Suppose we had both a minimum *and* a maxiumum size requirement. For example, what if our site had to be at least one acre but not larger than ten acres? Dissolving adjacent parcels might then disqualify some sites that were acceptable individually, like a five-acre site next to a six-acre site. We avoided this problem by not imposing a maximum size limit. We could have solved the problem procedurally, but we felt the solution was too involved for this book.

A second potential problem has to do with how attributes are managed in overlay operations. One of the later steps in the analysis uses an Identity overlay to assign demographic values from features in *BlockGroups* to the five recommended sites. Suppose that one of the sites lay across two block groups: which block group's attribute values would the site get? (Normally, this wouldn't happen because individual parcels don't cross block group boundaries. Dissolving the parcels, however, makes it a possibility.) If it happened, the geoprocessing tool would arbitrarily choose one block group or the other and assign its values to the site. We might not want an arbitrary solution, however. Instead, we might want to take the values from the block group that shared the most area with the site. Again, this problem can be solved procedurally, but not without effort. In our case, luckily, it didn't arise.

A third potential problem, inherent in any dissolve operation, is the loss of attributes in the feature class being dissolved. This wasn't an issue for us because we didn't need to maintain any parcel attributes except for ones, like ACRES, that we could calculate ourselves.

Selecting parcels by location

The first stage of the analysis focused on defining "good zones." These were areas that met our distance requirements with respect to the river and existing parks, and which also had suitable demographic values. After establishing these zones, we used a spatial query to select vacant parcels that fell within them. We chose the spatial operator "intersect" to select parcels that lay partially within a good zone; a containment operator would have selected only parcels that were entirely inside a good zone.

The main reason for our decision was simply to get more candidates. Our spatial query selected seven parcels (later reduced to five through aggregation). Had we used the "within" operator instead, four parcels would have been selected: the three that became Recommended Site 1, and the one that became Recommended Site 4. Sites 2 and 3 cross park buffer boundaries and Site 5 crosses the river buffer: these three sites wouldn't have been selected by a containment operator. Figure A-6 shows Site 3 as an example.

Figure A-6

Site 3 (white polygon) lies partially inside an orange "good zone." A small piece of it is within 0.25 miles of the park to the northwest (green polygon).

We have another reason to use the "intersect" operator. Site 3, in the figure above, is a 1.7-acre site which was dissolved from eight adjacent vacant parcels. It so happens that six of those eight parcels can be combined to make a 1.3-acre site that lies entirely within the good zone. Had we used a containment operator, Site 3 would not have been selected, even though a smaller desirable site is completely contained by the good zone at that location.

The problem with the "intersect" operator—if you consider it a problem—is that it selects a parcel even if only a small bit of it is within a good zone. Site 2, in Figure A-7, is an example.

Figure A-7

Only about 15 percent of Site 2's area overlaps a good zone.

Scoring

Our analysis had many siting requirements, but no way to rank candidates in terms of how well they met those requirements. We stated a preference for sites according to their proximity to the river and total population served, but these preferences weren't quantified in any way. The final evaluation of candidates may be too subjective and multifaceted to make strict scoring possible; nevertheless, we can briefly consider a couple of techniques.

For a variable of interest, you could map values onto a standardized score from zero to one by assigning 1 to the best value and 0 to the worst value and calculating other values proportionally. Using distance to the river as an example, you'd get the scores in the table below:

Site	Distance to LA River (feet)	Standardized score (reversed)
1	3,086	0.15
2	2,482	0.33
3	191	1
4	2,537	0.31
5	3,603	0

Scores are reversed because a low value is desirable.

Assuming all variables are equally important and can be standardized with the same linear function, you could rank the sites by scoring all variables and then summing the scores. (Ultimately, this approach needs normalization or else variables with small value ranges are differentiated as much as those with large value ranges.)

Your values may not be equally important, of course. A site's proximity to the river might matter more than the percentage of children in the neighborhood. This leads to the subject of multicriteria evaluation methods and the assignment of weights (multipliers) to variables. We won't pursue this subject here except to note that ArcGIS provides, in the raster analysis environment, a tool called Weighted Overlay to facilitate this kind of scoring.

Other factors

An analysis can be refined and developed to the extent that data, ingenuity, and time allow. In addition to the topics discussed above, we made simplifying assumptions and omitted several factors that might be important in choosing a park. For example:

- Land ownership. (We assumed that all vacant land was available. A refinement might consider only city-owned vacant parcels.)
- Land acquisition and development cost.
- Neighborhood crime rates.
- Environmental quality of the site. (This could include soil toxicity, proximity to freeways, and other health risks.)
- Accessibility from surface streets.
- Aesthetic qualities. (These might include visibility of the river, presence of trees, eyesores like overhead power lines, ambient noise levels, and so on.)

B Data and photo credits and data license agreement

Data and photo credits

Cover

Imagery sources: USGS, FAO, NPS, EPA, Esri, DeLorme, TANA, other suppliers.

Lesson 1

Page 3 photo credit: Security Pacific National Bank Collection, Los Angeles Public Library, Anthony Friedkin, Photographer.

Imagery sources: Streets Basemap–Esri, DeLorme, AND, Tele Atlas, First American, Esri Japan, UNEP-WCMC, USGS, METI, Esri Hong Kong, Esri Thailand, Procalculo Prosis; Imagery basemap–Esri, i-cubed, USDA FSA, USGS, AEX, GeoEye, Getmapping, Aerogrid, IGP.

Data sources: U.S. Populated Place Areas from Esri Data & Maps 2009, courtesy of Tele Atlas, U.S. Census; River from Esri Data & Maps 2009, courtesy of USGS, Esri; U.S. and Canada Parks from Esri Data & Maps 2009, courtesy of Tele Atlas; U.S. Tracts from Esri Data & Maps 2009, courtesy of Tele Atlas, U.S. Census, Esri (Pop2007 field); U.S. Block Groups from Esri Data & Maps 2009, courtesy of Tele Atlas, U.S. Census, Esri; LA City Land Parcels–courtesy of LA City, Public Works, Bureau of Engineering.

Lesson 2

Imagery sources: Imagery basemap–Esri, i-cubed, USDA FSA, USGS, AEX, GeoEye, Getmapping, Aerogrid, IGP.

Data sources: LA City Land Parcels–courtesy of LA City, Public Works, Bureau of Engineering; U.S. Populated Place Points from Esri Data & Maps 2009, courtesy of U.S. Census; Parks.shp–courtesy of LA City, Public Works, Bureau of Engineering; U.S. Block Groups from Esri Data & Maps 2009, courtesy of Tele Atlas, U.S. Census, Esri; U.S. Block Centroids–Esri Data & Maps 2009; U.S. Tracts from Esri Data & Maps 2009, courtesy of Tele Atlas, U.S. Census, Esri (Pop2007 field); LA City Vacant Parcels table–courtesy of LA City, Public Works, Bureau of Engineering.

Lesson 3

Imagery sources: Physical Basemap–USGS, NPS, Esri, TANA, AND; Imagery Basemap–Esri, i-cubed, USDA FSA, USGS, AEX, GeoEye, Getmapping, Aerogrid, IGP.

Data sources: LARiver Centerline–courtesy of LA City, Public Works, Bureau of Engineering; U.S. and Canada Water Polygons from Esri Data & Maps 2009, courtesy of Tele Atlas; River from Esri Data & Maps 2009, courtesy of USGS, Esri; U.S. Populated Place Areas from Esri Data & Maps 2009, courtesy of Tele Atlas, U.S. Census; Parks.shp–courtesy of LA City, Public Works, Bureau of Engineering; U.S. and Canada Parks from Esri Data & Maps 2009, courtesy of Tele Atlas; VistaHermosaPark.tif–courtesy of Aerials Express from DataDoors; U.S. Counties from Esri Data & Maps 2009, courtesy of ArcUSA, U.S. Census, Esri (Pop2009 field); U.S. States from Esri Data & Maps 2009, courtesy of Tele Atlas; Continents from Esri Data & Maps 2009, courtesy of ArcWorld Supplement; LA City Vacant Parcels table–courtesy of LA City, Public Works, Bureau of Engineering.

Lesson 4

Data sources: LA City Land parcels–courtesy of LA City, Public Works, Bureau of Engineering; U.S. Highways from Esri Data & Maps 2009, courtesy of Tele Atlas; U.S. Counties from Esri Data & Maps 2009, courtesy of ArcUSA, U.S. Census, Esri (Pop2009 field); U.S. Populated Place Areas from Esri Data & Maps 2009, courtesy of Tele Atlas, U.S. Census; U.S. and Canada Parks from Esri Data & Maps 2009, courtesy of Tele Atlas; U.S. Block Groups from Esri Data & Maps 2009, courtesy of Tele Atlas, U.S. Census, Esri; LA City Vacant Parcels table–courtesy of LA City, Public Works, Bureau of Engineering.

Lesson 5

Imagery sources: Imagery Basemap–Esri, i-cubed, USDA FSA, USGS, AEX, GeoEye, Getmapping, Aerogrid, IGP.

Data sources: U.S. and Canada Parks from Esri Data & Maps 2009, courtesy of Tele Atlas.

Lesson 6

Site photos on pages 233-235 by Christian Harder.

Imagery sources: Topographic Basemap–USGS, FAO, NPS, EPA, Esri, DeLorme, TANA, other suppliers; Imagery Basemap–Esri, i-cubed, USDA FSA, USGS, AEX, GeoEye, Getmapping, Aerogrid, IGP.

Data sources: River from Esri Data & Maps 2009, courtesy of USGS, Esri; U.S. Populated Place Points from Esri Data & Maps 2009, courtesy of U.S. Census; U.S. Counties from Esri Data &

Maps 2009, courtesy of ArcUSA, U.S. Census, Esri (Pop2009 field); LARiver Centerline–courtesy of LA City, Public Works, Bureau of Engineering; U.S. and Canada Parks from Esri Data & Maps 2009, courtesy of Tele Atlas; LA City Land parcels–courtesy of LA City, Public Works, Bureau of Engineering; U.S. Block Groups from Esri Data & Maps 2009, courtesy of Tele Atlas, U.S. Census, Esri; U.S. Block Centroids–Esri Data & Maps 2009; U.S. Large Area Landmarks from Esri Data & Maps 2009, courtesy of Tele Atlas; LA City Vacant Parcels table–courtesy of LA City, Public Works, Bureau of Engineering.

Lesson 7

Data sources: LARiver Centerline–courtesy of LA City, Public Works, Bureau of Engineering; LA City Vacant Parcels table–courtesy of LA City, Public Works, Bureau of Engineering; U.S. Block Centroids–Esri Data & Maps 2009; U.S. Block Groups from Esri Data & Maps 2009, courtesy of Tele Atlas, U.S. Census, Esri; U.S. and Canada Parks from Esri Data & Maps 2009, courtesy of Tele Atlas.

The pancake recipe is thanks to Prof. Emeritus, Dr. Tech. Ole Jacobi, Institute of Surveying and Photogrammetry, Technical University of Denmark.

Lesson 8

Imagery sources: Imagery Basemap–Esri, i-cubed, USDA FSA, USGS, AEX, GeoEye, Getmapping, Aerogrid, IGP; Terrain Basemap–USGS, Esri, TANA, AND.

Data sources: LARiver Centerline–courtesy of LA City, Public Works, Bureau of Engineering; LA City Land parcels–LA City, Public Works, Bureau of Engineering; U.S. Populated Place Areas from Esri Data & Maps 2009, courtesy of Tele Atlas, U.S. Census; U.S. Counties from Esri Data & Maps 2009, courtesy of ArcUSA, U.S. Census, Esri; U.S. Highways from Esri Data & Maps 2009, courtesy of Tele Atlas.

Lesson 9 (online)

Imagery sources: Imagery Basemap–Esri, i-cubed, USDA FSA, USGS, AEX, GeoEye, Getmapping, Aerogrid, IGP.

Data Sources: River from Esri Data & Maps 2009, courtesy of USGS, Esri; U.S. Median Household Income: © 2010 Esri.

Appendix A

Imagery sources: Imagery Basemap–Esri, i-cubed, USDA FSA, USGS, AEX, GeoEye, Getmapping, Aerogrid, IGP.

Data sources: U.S. Block Groups from Esri Data & Maps 2009–courtesy of Tele Atlas, U.S. Census, Esri; U.S. Tracts from Esri Data & Maps 2009–courtesy of Tele Atlas, U.S. Census, Esri.

Data license agreement

Important: Read carefully before opening the sealed media package

Environmental Systems Research Institute, Inc. (Esri), is willing to license the enclosed data and related materials to you only upon the condition that you accept all of the terms and conditions contained in this license agreement. Please read the terms and conditions carefully before opening the sealed media package. By opening the sealed media package, you are indicating your acceptance of the Esri License Agreement. If you do not agree to the terms and conditions as stated, then Esri is unwilling to license the data and related materials to you. In such event, you should return the media package with the seal unbroken and all other components to Esri.

Esri License Agreement

This is a license agreement, and not an agreement for sale, between you (Licensee) and Environmental Systems Research Institute, Inc. (Esri). This Esri License Agreement (Agreement) gives Licensee certain limited rights to use the data and related materials (Data and Related Materials). All rights not specifically granted in this Agreement are reserved to Esri and its Licensors.

Reservation of Ownership and Grant of License:

Esri and its Licensors retain exclusive rights, title, and ownership to the copy of the Data and Related Materials licensed under this Agreement and, hereby, grant to Licensee a personal, nonexclusive, nontransferable, royalty-free, worldwide license to use the Data and Related Materials based on the terms and conditions of this Agreement. Licensee agrees to use reasonable effort to protect the Data and Related Materials from unauthorized use, reproduction, distribution, or publication.

Proprietary Rights and Copyright:

Licensee acknowledges that the Data and Related Materials are proprietary and confidential property of Esri and its Licensors and are protected by United States copyright laws and applicable international copyright treaties and/or conventions.

Permitted Uses:

Licensee may install the Data and Related Materials onto permanent storage device(s) for Licensee's own internal use.

Licensee may make only one (1) copy of the original Data and Related Materials for archival purposes during the term of this Agreement unless the right to make additional copies is granted to Licensee in writing by Esri.

Licensee may internally use the Data and Related Materials provided by Esri for the stated purpose of GIS training and education.

Uses Not Permitted:

Licensee shall not sell, rent, lease, sublicense, lend, assign, time-share, or transfer, in whole or in part, or provide unlicensed Third Parties access to the Data and Related Materials or portions of the Data and Related Materials, any updates, or Licensee's rights under this Agreement.

Licensee shall not remove or obscure any copyright or trademark notices of Esri or its Licensors.

Term and Termination:

The license granted to Licensee by this Agreement shall commence upon the acceptance of this Agreement and shall continue until such time that Licensee elects in writing to discontinue use of the Data or Related Materials and terminates this Agreement. The Agreement shall automatically terminate without notice if Licensee fails to comply with any provision of this Agreement. Licensee shall then return to Esri the Data and Related Materials. The parties hereby agree that all provisions that operate to protect the rights of Esri and its Licensors shall remain in force should breach occur.

Disclaimer of Warranty:

The Data and Related Materials contained herein are provided "as-is," without warranty of any kind, either express or implied, including, but not limited to, the implied warranties of merchantability, fitness for a particular purpose, or noninfringement. Esri does not warrant that the Data and Related Materials will meet Licensee's needs or expectations, that the use of the Data and Related Materials will be uninterrupted, or that all nonconformities, defects, or errors can or will be corrected. Esri is not inviting reliance on the Data or Related Materials for commercial planning or analysis purposes, and Licensee should always check actual data.

Data Disclaimer:

The Data used herein has been derived from actual spatial or tabular information. In some cases, Esri has manipulated and applied certain assumptions, analyses, and opinions to the Data solely for educational training purposes. Assumptions, analyses, opinions applied, and actual outcomes may vary. Again, Esri is not inviting reliance on this Data, and the Licensee should always verify actual Data and exercise their own professional judgment when interpreting any outcomes.

Limitation of Liability:

Esri shall not be liable for direct, indirect, special, incidental, or consequential damages related to Licensee's use of the Data and Related Materials, even if Esri is advised of the possibility of such damage.

No Implied Waivers:

No failure or delay by Esri or its Licensors in enforcing any right or remedy under this Agreement shall be construed as a waiver of any future or other exercise of such right or remedy by Esri or its Licensors.

Order for Precedence:

Any conflict between the terms of this Agreement and any FAR, DFAR, purchase order, or other terms shall be resolved in favor of the terms expressed in this Agreement, subject to the government's minimum rights unless agreed otherwise.

Export Regulation:

Licensee acknowledges that this Agreement and the performance thereof are subject to compliance with any and all applicable United States laws, regulations, or orders relating to the export of data thereto. Licensee agrees to comply with all laws, regulations, and orders of the United States in regard to any export of such technical data.

Severability:

If any provision(s) of this Agreement shall be held to be invalid, illegal, or unenforceable by a court or other tribunal of competent jurisdiction, the validity, legality, and enforceability of the remaining provisions shall not in any way be affected or impaired thereby.

Governing Law:

This Agreement, entered into in the County of San Bernardino, shall be construed and enforced in accordance with and be governed by the laws of the United States of America and the State of California without reference to conflict of laws principles. The parties hereby consent to the personal jurisdiction of the courts of this county and waive their rights to change venue.

Entire Agreement:

The parties agree that this Agreement constitutes the sole and entire agreement of the parties as to the matter set forth herein and supersedes any previous agreements, understandings, and arrangements between the parties relating hereto.

Index

colors, customizing 294
coordinate systems. *See also* on-the-fly projection; projecting data
  choosing 122
  and datums 116
  described 102
  in folders 109, 173
  geographic 102, 104, 110
  and map projections 111
  projected 102, 107, 111
  State Plane 113
copying features 135–136, 260–261
Create Features window 174, 183

D

data
  acquiring 64
  exporting 208, 211
  field types 144
  formats. *See* geodatabases; shapefiles
  loading 142–145
  overwriting 148
  quality of 169, 339
  raster 63, 72, 340
  storage precision 144
  vector 63, 340
databases. *See* geodatabases
data frame properties, opening 106, 303–304
data frames
  active 106, 303
  borders around 304
  coordinate systems in 109, 113–115, 172–173
  copying 307
  described 105, 288
  focusing 320, 324
  inserting 106, 303
  renaming 300, 307
  resizing 303–304
data models 63, 340
data requirements table 59
datasets, compared to layers 17, 91
data types. *See* field data types
datums 103, 116, 127. *See also* geographic transformations
dBASE format 63, 67
default geodatabases 93, 121, 195
definition queries 15–17, 18–19
degrees, notation of 105
digital elevation models (DEM) 72

dissolving features 140–142, 164–165, 343
distance analysis 196–197, 341–342
distances
  calculating 219–220
  measuring 44–45
division by zero 157
dynamic text 325, 327

E

editing
  attribute values 184–186
  described 169
  existing features 175–180
  new features 184–185
  shortcut keys 176
  snapping behavior in 179
  warnings 172
  in workspaces 188
edit sessions
  starting 172, 183
  stopping 180
edit sketches 176, 179, 181
environment settings 129, 208, 279
erasing features 197, 200, 248–250
exporting data 208, 211
exporting maps 53
extent, fixing 299

F

feature classes
  converting to geodatabase 130, 149–151
  deleting 147, 167, 168, 230
  described 63
  importing 149–151
  renaming 131, 167
feature datasets 63, 64, 122
feature identifiers. *See* FID; OBJECTID
features. *See also* feature classes, records
  clearing selected 15
  described 11
  geometry types 63
  identifying 13, 18, 66–67, 71, 95
  multipart 137, 140, 165, 199
  selecting 21, 133, 137
  subselecting 210
  zooming to selected 15
feature templates 183–184
FID 69, 70
field data types 144

fields. *See also* attributes
  adding 152
  aliasing 150, 152, 227
  calculating 154–155, 156–157, 188, 226. *See also* calculating geometry
  deleting 225–226
  formatting 227–228
  key. *See* table joins
  naming 152
  sorting 14, 87
  statistics for 39–40, 46, 140
  turning off and on 134–135
folders
  connecting to 10
  coordinate systems in 109, 173
  creating 121
  home 27, 32, 93
  My Documents\ArcGIS 22, 121

G

geodatabase annotation 327
geodatabases. *See also* feature classes
  compared to shapefiles 63, 152, 158
  creating 121, 194, 239
  default 93, 121, 195
  planning 122
  types of 123
geographic coordinate systems. *See also* datums
  described 102, 104, 110
  NAD 1983 114, 127
  WGS 1984 103–104, 130
geographic coordinate systems warnings 90, 114–115, 116, 132. *See also* geographic transformations
geographic transformations 116, 127, 129, 173
geoprocessing. *See also* geoprocessing tools
  adding results of to display 269, 270
  environment settings in 129, 208, 279
  options 128, 148, 270
  overwriting outputs of 148, 269, 270
  results 128, 279
geoprocessing tools
  Aggregate Polygons 213–215
  Buffer 195, 196, 198–199, 216–217, 244–248
  Copy Features 135–136, 260–261
  Create Thiessen Polygons 196
  Dissolve 139–141, 164–165
  Erase 197, 200, 248–250
  Feature Class to Feature Class 149–151
  Feature Class to Geodatabase (multiple) 130

geoprocessing tools (*continued*)
  Identity 197, 204, 221, 222, 251–252
  Intersect 197
  Make Feature Layer 223, 255, 256, 258–259
  Minimum Bounding Geometry 342
  Near 197, 219–220
  Project 125–127
  Select Layer By Attribute 255–257
  Select Layer By Location 259–260
  Spatial Join 197, 218–219, 262–263
  Union 197
  Upgrade Metadata 78
  Weighted Overlay 345
geoTIFF file format 53, 232
gradient fills 291–292
graphics. *See also* map annotation
  converting to features 232
  creating 50–52, 323–324
  described 51
  rotating 51
  symbolizing 50, 323
graphic text 324–325, 327, 330–331
graphs 309–315
grids 288–289
group layers 24
guides 314

H

halos 36
help
  ArcGIS Desktop 95, 125, 265
  tool 125, 266
histograms
  in data classification 41–42
  in field statistics 39
home folder 27, 32, 93

I

identifying features 13, 18, 66–67, 71, 95
identity overlays 197, 204, 221, 222, 251–252, 343
indexes
  for attributes 162–163
  for maps, data, and tools 139
inset maps 283, 303
intermediate data
  deleting 230
  described 122
  in models 251, 256, 265, 269
  naming 195

tracts. *See* census geography
transparency, setting 43, 94, 293

U

undoing operations 169, 177, 185, 274
unprojected data 104

V

vector data 63, 340
vertices
  deleting 177
  described 181
  editing 175–180
  modifying 177–178
viewer windows 106–107

W

WGS 1984 103–104, 130
workflow. *See* analysis: planning
workspaces 188

Z

zooming
  to extents 34, 97, 112, 179
  to features 15, 95, 211
  interactively 9
  to layers 14
  in layout view 290, 324
  in models 257
  in previews 65

Topics and call-out boxes